SHANG-KENG MA was born in Chungking, China on September 24, 1940. After completing his high school education in Taiwan, he undertook higher studies at the University of California, Berkeley where he earned his PhD in 1966. In 1975 he became professor of physics at the University of California, San Diego.

Ma is perhaps best known for his contributions to the theory of critical phenomena. In collaboration with Halperin and Hohenberg he pioneered the application of renormalization group theory to critical dynamics. His work with Nickel and Fisher on the effect of long-range interactions is widely known. In the last ten years of his life, he was instrumental in the development of the Monte Carlo renormalization group theory, the random magnetic Ising model, the dynamics of spin glasses and the "coincidence counting method."

Despite his declining health he continued to teach until a few days before his death. He died in his home in La Jolla on November 24, 1983.

FROM THE PREFACE: In finding for this book a theme appropriate to Shang-keng Ma's memory, we were guided by the idea that he was interested in essentially all areas of theoretical physics. He would give any subject a fair hearing in selecting those bits of fields that most appealed to him. Since the bulk (though not all) of his work was in condensed matter physics, we decided to try to assemble a collection of review-style articles in active, emerging areas of condensed matter theory. Feeling that we would achieve the most stimulating and topical book by encouraging the authors to write about subjects which intrigue them most at present, we made no attempt to constrain the distinguished contributors as to topic. We told them only that we wanted the volume to be irresistible to a graduate student looking for an exciting area in which to begin research. We are delighted by all of the contributions, and grateful to the authors for their considerable efforts and patience. In the final analysis we feel pleased with this collection because we can imagine Shang wanting to pick it up and read it.

World Scientific Series on Directions in Condensed Matter Physics — Vol. 1

DIRECTIONS IN CONDENSED MATTER PHYSICS

Memorial Volume in Honor of Shang-keng Ma

Edited by

G. Grinstein
G. Mazenko

World Scientific

Published by

World Scientific Publishing Co Pte Ltd.
P. O. Box 128, Farrer Road, Singapore 9128
242 Cherry Street, Philadelphia PA 19106-1906, USA

Library of Congress Cataloging-in-Publication Data

Directions in condensed matter physics.

   1. Condensed matter.   2. Ma, Shang-keng, 1940–1983.
I. Ma, Shang-keng, 1940–1983.   II. Grinstein, Geoffrey.
III. Mazenko, Gene.
QC173.4.C65D57   1986   530.4'1   86–18920
ISBN 9971-978-42-3
ISBN 9971-978-58-X (pbk.)

DIRECTIONS IN CONDENSED MATTER PHYSICS
Memorial Volume in Honor of Shang-keng Ma

Printed in Singapore by Fu Loong Lithographer Pte Ltd.

PREFACE

This volume is dedicated to the memory of our colleague, teacher, and friend, Shang-keng Ma, who died at age 43 in November, 1983, after a protracted battle with cancer. The premature death of this kind, gifted, incredibly modest man has left a painful void in the lives of those of us who worked with him.

We met Shang shortly after completing our graduate studies, and soon entered into collaboration with him. In the rare moments not spent fretting over job prospects, most of our contemporaries in physics at the time (we certainly among them) were casting about for a salutary style of living their lives as physicists. Among the many forceful and charismatic personalities of theoretical physics, the young, mild and self-effacing Shang Ma seemed an unlikely role model for this search. It was only in working with him that one began to understand his special, to us irresistibly appealing, approach to doing physics.

When we first started working with Shang there was excitement among condensed matter theorists over the newly-discovered renormalization-group techniques for treating critical phenomena. He had already contributed weightily to the development of these ideas through, for example, his work on the $1/n$ expansion, his elucidation, with Nickel and Fisher, of the effect of long-range interactions, and his invention, with Halperin and Hohenberg, of renormalization-group methods for critical dynamics. Compared to his knowledge and achievements in this field, ours were minimal. This would have been impossible to infer from the tone of our discussions, however. Shang had an instinct for equality and an instinctive sympathy for the underdog; the lowliest student and the most radiant luminary received at his hands the same dignified courtesy. While he could be keenly critical of ideas, evaluating people was distasteful to him. In his presence one felt secure that one was not on trial. One's mere interest in a problem that excited him was sufficient recommendation. Trying to make finer judgments was considered a waste of time, and anyone with enough energy was assumed capable of contributing to the general understanding.

It is hard to overestimate how liberating it was for a young person to work in such an atmosphere. One felt protected from the relentless competition of the world outside. The problem under study was paramount, and there was quiet confidence that the problem could be solved; worry about "being scooped" and "having impact" seemed petty. In the face of Shang's generosity of spirit (his version, having nothing to do with fact, was always that his co-workers were responsible for 100% of any progress achieved, his role being something along the lines of keeping them supplied with sharp pencils), small-minded impulses in his collaborators withered and died.

Much of Shang's early work involved field theoretic pyrotechnics, and many of us who worked with him fairly early on still harbor the suspicion that he had occult powers of diagrammatic analysis. In the most tangled mass of Feynman graphs, Shang somehow knew where the interesting behavior lay. Sometimes he even had difficulty explaining how he knew, though days of labor by the rest of us would usually prove him right. We remember one occasion late in 1975 when the three of us, cloistered in a small office on the first floor of the physics building in La Jolla, were trying to extract the dominant infrared singularity from a disheartening set of diagrams. Though straightforward, the analysis was tedious. We two sat at our desks, methodically depleting the world's proven pulp and paper reserves. Shang sat on a chair in the corner, frowning at the proverbial back of the envelope — in this case a small airmail envelope with little space for mistakes. Every half hour or so he would make an invisible mark on it with a fountain pen. Then he would cross the mark out, looking ever more glum. After some hours of this the two of us with desks and paper had arrived at results. Neither had any confidence in his answer; the two results disagreed violently. We turned to Shang. He gave us his envelope. In the center of the scratchings was a minute algebraic expression, which agreed with one of ours (there seems now to be dispute over precisely whose) and turned out to be correct. There were no supporting calculations. Much impressed by this, we asked him to explain how he had done it. He began in his usual matter-of-fact way, unclear as to what the fuss was about, but could see that we weren't following. He stopped and pondered a bit. Suddenly his face was alight with the key to elucidating the method: "When you think about it a little," he said, "what else could it be?"

"Anything else!" we thought, but were so awed by the mystic timbre of his question that we stood there dumbly, pretending all was clear. Shang seemed pleased. A moment later we were interrupted, and never did get a chance to press him for a less ancient explanation.

As his career progressed, Shang came to rely less on formal field-theoretic calculations and more on simpler physical arguments. One could already sense this trend in the mid-70's, when much of his work still involved diagrammatic analysis. (This was a marvellously productive period for Shang. Between the years 1972 and 1979 he developed (with Halperin and Hohenberg) the dynamical renormalization group mentioned earlier, and was the first (together with Mazenko) to incorporate mode-coupling terms in this framework. He discovered, in superconductors (with Halperin and Lubensky), the first example of the important phenomenon now known as "fluctuation-induced first-order phase transitions," produced (with Imry) the classic paper on random fields, and invented the Monte Carlo renormalization group. His book on critical phenomena, published in 1976, remains a standard reference in the field.)

We caught a wonderful glimpse of Shang's gradually shifting focus on the morning following the incident recounted above. Having assumed our places we two were having difficulty concentrating on the task at hand. (In fact, we recall embarking on a less-than-scholarly debate over which group performed the song "Give Me Just a Little More Time.") Shang sat motionless in his corner with a half sheet of paper. So far as we remember he wrote nothing, again looking more and more miserable as the morning wore on. We stopped for lunch. It was clear that Shang, who usually seemed so phlegmatic, was dejected. We asked what the problem was. "Oh," he said, "I've wasted the entire morning. I couldn't even do the simplest calculation."

"What were you trying to calculate?"

"Oh nothing. I was just trying to estimate the sign of some quantity."

We tried to control ourselves, but the idea of trying to "estimate" a sign struck us as so funny that we were soon choking with the effort. A pale smile broke through the gloom in Shang's expression, and at once we were all howling with laughter that made conversation elsewhere in the canteen impossible. (Later that afternoon a relieved Shang told us that the discovery of a missing factor of -1 had improved his estimate, and that the sign was looking more encouraging.) "Estimating the sign" became a standing joke among us, but "estimate" continued to replace "calculate" in Shang's subsequent descriptions of the physics he was doing. There was no diminution of his mathematical facility. More and more often, however, the questions he found most absorbing were those for which an insight into the physics purchased more than a calculation. This evolution persisted right up until his death, and is discernible in his later work, notably in his invention of an extremely original coincidence counting method for computing the entropy of complex systems, and his important contributions to the understanding of the random field Ising model. His book on statistical mechanics (written in Chinese and only recently translated into English) is full of ingenious physical arguments.

The sense of security one had working with Shang was not entirely a result of his largeness of spirit. It derived in part from one's experience that when the problem in question turned nasty Shang would find a window through which it looked benign. His way of seeing a problem was seldom the same as other people's, and seldom constant for four days running. He was ruthless about discarding an approach in which flaws began to appear, no matter how much prior effort it represented. He was forever turning things around, standing them on their ears, trying to make them yodel. In a collaboration with Shang one risked being uprooted every morning. Overnight the whole program might have been scrapped and a more promising one devised. Those of us more senti- mental about our pasts than he were often left with a feeling of weightlessness — struggling for balance, straining to keep up, starting from scratch *again*. Eventually a satisfactory course would be found and we would have a few hours of peaceful calculating. Then off we would go again, destination only vaguely known.

Occasionally Shang's thinking about a problem was so unconventional as to be hard to follow. Most of us who worked with him had the experience of dismissing as nonsense a particularly alien-sounding idea of his, only to rediscover it for ourselves after weeks of work. Shang seemed unconcerned about being misunderstood in this way. He was decidedly lacking in evangelistic fervor. It seemed sufficient for him to have convinced himself of the validity of a point of view. Others would eventually come around if his reasoning were sound.

One encountered a similar self-sufficiency when one tried to explain things to him. He was always supportive, and open to new ideas, but when one grew to know him one could perceive a sharp distinction between ideas that appealed to him and ones that simply did not. One was sometimes drawn to undignified excesses of salesmanship in trying to promote a favorite thought from the latter category to the former. Occasionally one sank so low as to dredge up a scrap of evidence to show that one of the great men in the field subscribed to the doctrine one was espousing. This never produced any response but a shy laugh, though it may from time to time have stiffened Shang's resistance. At such moments one could feel the fierceness of his intellectual independence. He had to understand everything in his own way. Lore which did not make sense to him left him suspicious, no matter how common its acceptance or illustrious its origins. He was not the least intimidated by the views of celebrated physicists. Physics was not a spectator sport to Shang, and he saw the hero worship which so strangely pervades the profession as an impediment to his participation.

While not indifferent to others' opinions of his abilities, Shang refused to grant much importance to such judgments. He never hesitated to ask a question that might further his understanding, even if that question exposed a gap in his knowledge. We have all been in a group where someone uses the latest "buzz" word and everyone agrees knowingly that "buzz" is just great. There is a tacit conspiracy that no one will demand to know what "buzz" means. In such situations Shang would ask, without aggression, what the rest of us lacked the confidence to ask: "What is 'buzz,' anyhow?" (Once, as a session chairman at a March meeting of the American Physical Society he delighted the crowd by asking the author of a paper with a five-letter acronym in its title what the initials stood for.) One then experienced the suspense of discovering whether anyone could in fact explain "buzz." It was on such occasions that we loved Shang best, for his honesty and courage.

Quiet courage remained a part of Shang's make-up to the end of his life. He continued to carry out his teaching duties at UCSD despite the severe pain and debilitation of his disease. Scant days before his death he consulted a doctor who pronounced him gravely ill, at which point he felt that he had satisfactorily discharged his responsibilities and could stop teaching. It was characteristic of Shang's independence of mind that this was the first time he had seen a doctor in many months. Having concluded that contemporary medicine could do nothing further for him, he had terminated conventional medical treatment to try to cope with his illness in his own way.

Shang-keng Ma adhered as determinedly as anyone we have met to the vision which draws so many of us to theoretical physics in the first place: the image of sitting in a quiet spot with pencil, paper, and perhaps a friend, and thinking hard about a problem until one understands something about it. Through the years that intensely personal vision tends to get lost in a swarm of job applications, priority squabbles, funding anxieties, jet lags, learned committees, lustful thoughts of honors and prizes, and the rest of the ogres in the punishing road of "career advancement." In hindsight it seems comically naive; one forgets its original power. Shang managed never to take his eyes off that youthful ideal. He paid dues, like the rest of us, to the establishment which supported him, but always remained slightly aloof from it, refusing to let the less noble of its values clutter his path. For all his mildness he was absolutely uncompromising on this point. There was nothing of the huckster in Shang. The idea of promoting himself or his work was utterly foreign to him. He sacrificed to these principles some recognition for his scientific accomplishments. We feel certain that the sacrifice was made consciously and without regret. For those of us lucky enough to have known him, he leaves behind more than a significant body of research. He leaves behind a standard against which we measure our conduct as physicists and as human beings. Despite our sadness in being unable to call Shang, to gossip with him and to hear about his latest ideas, there is something comforting in the picture we carry of his gentle, sympathetic smile — something that helps us remember why we are doing what we do.

In finding for this book a theme appropriate to his memory, we were guided by the idea that Shang-keng Ma was interested in essentially all areas of theoretical physics. He would give any subject a fair hearing in selecting those bits of fields that most appealed to him. Since the bulk (though not all) of his work was in condensed matter physics, we decided to try to assemble a collection of review-style articles in active, emerging areas of condensed matter theory. Feeling that we would achieve the most stimulating and topical book by encouraging the authors to write about subjects which intrigue them most at present, we made no attempt to constrain the distinguished contributors as to topic. We told them only that we wanted the volume to be irresistible to a graduate student looking for an exciting area in which to begin research. We are delighted by all of the contributions, and grateful to the authors for their considerable efforts and patience. In the final analysis we feel pleased with this collection because we can imagine Shang wanting to pick it up and read it.

Yorktown Heights, N. Y.   Geoffrey Grinstein
Chicago, Illinois   Gene Mazenko

CONTENTS

DIRECTIONS IN CONDENSED MATTER PHYSICS
Memorial Volume in Honor of Shang-keng Ma

PERCOLATION

Amnon Aharony

School of Physics and Astronomy
Raymond and Beverly Sackler Faculty of Exact Sciences
Tel Aviv University, Tel Aviv 69978
ISRAEL

1. Introduction

The mathematical problem of *percolation* is very easily defined: consider an infinite periodic lattice of sites (or bonds). Each site (or bond) is occupied (with probability p) or empty (with probability $1 - p$). Nearest neighbor occupied sites (bonds) form connected *clusters*. As p increases from zero, the average size of these clusters increases. For the infinite system, this average size diverges at the *percolation threshold*, p_c. Above p_c the system contains an *infinite* cluster, which connects between its edges.

Near p_c, the only length scale relevant for describing the properties of the system is the *connectedness length*, ξ, defined as the root mean square distance between pairs of sites (or bonds) which belong to the same finite cluster. Since ξ diverges to infinity at p_c, there remains no measuring unit with which to scale the dependence of various properties on lengths. As a result, the system looks qualitatively the same when looked at with different magnifications. This phenomenon, called *self-similarity,* is responsible for various anomalous properties of dilute systems near their percolation threshold. Since these properties arise at *all* length scales, they do not depend on many local details (e.g., the lattice structure), which become *irrelevant* to the large scale behavior. This yields many *universal* quantities, which depend only on the *dimensionality* of the system. It is thus sufficient to find these quantities on simple models, and then apply them to a wide variety of realistic situations.

The existence of the infinite cluster only above p_c is similar to the appearance of an order parameter in a phase transition. The divergence of the connectedness length at p_c also resembles that of the correlation length near critical points. Indeed, all the theoretical modern tools of critical phenomena can be used to describe this geometrical transition. Some of these are reviewed below, in Secs. 2 and 3. These sections form an attempt to present the theory in a didactic way, with a strong emphasis on the *geometry*.

Obviously, percolation theory is relevant to many physical dilute or alloyed systems. It is impossible to do justice to all the interesting applications in this review. Instead, I chose to present two specific examples, which exhibit the strong influence of the self-similarity geometry on the physics. Section 4 discussed dilute magnets, and Sec. 5 reviews some electrical conductivity and diffusion phenomena. Naturally, the choice of the examples included here was affected by my personal taste. This fact, and the attempt to be didactic, must have caused the absence of many relevant references, and I apologise to their authors. Several excellent recent reviews contain much more complete lists of references, and I refer the readers to those for more details (Essam, 1980; Stauffer, 1979, 1985; Deutscher, Zallen and Adler, 1983; Stinchcombe, 1986 and references therein).

2. Fractal Geometry and Scaling

2.1. Self-similarity and fractal dimensionality

It is easy to produce by computer a percolation picture. One simply runs through each lattice site (or bond), and decides to occupy it (or not) using random numbers between 0 and 1. There exist various algorithms to identify connected clusters, and some of these were reviewed in detail by Stauffer (1979, 1985). Given a finite sample realization, one can perform various measurements of geometrical properties of the clusters, and collect data on their statistics.

Figure 1a shows a computer simulation of site percolation on the triangular lattice, at its threshold concentration $p_c = \frac{1}{2}$. The sites on the largest cluster, which connect between the boundaries of the finite sample, are emphasized by showing the bonds which connect them. Figure 1b then shows a coarse graining of the same picture: sites on the lattice were grouped into cells of three sites (forming triangles). Each cell was then replaced by an occupied single new site if a majority of its sites was occupied, and by an empty new site otherwise. Qualitatively, Fig. 1b cannot be distinguished from a piece of Fig. 1a. It is impossible to tell from the picture at what level of coarse graining, or magnification, the two pictures were taken. This is a qualitative manifestation of *self-similarity*.

Quantitatively, we may verify self-similarity by several measurements. First, we may choose a point on the large cluster (e.g., at the center of the picture) and count the sites on the cluster within a box of linear size L. For large L, the average of this number approaches an *asymptotic* power law,

$$M(L) \approx \overline{A} L^D \quad . \tag{2.1}$$

This behavior is to be contrasted with that of a homogeneous system, in which $M(L) = \rho L^d$, where d is the Euclidean dimensionality ($d = 2$ in Fig. 1) and ρ is the uniform density. The exponent D, which is found to be close to 1.9 at $d = 2$, to 2.5 at $d = 3$, and equal to 4 for $d > 6$, is called the *fractal dimensionality* (Mandelbrot, 1982) of the cluster. Power laws like (2.1) are expected whenever there exists no other length by which L can be scaled.

An alternative measurement of D arises from a comparison of the number of sites on the largest clusters of Figs. 1a and 1b. Since the unit of length was changed by a factor $b = \sqrt{3}$, the length L in Fig. 1a becomes L/b in Fig. 1b. If there exists no other parameter on which $M(L)$ depends, then $M(L)$ must be a *homogeneous function*, of the form

$$M(L) = b^D M(L/b) \quad , \tag{2.2}$$

The power law $M(L) \sim L^D$ is the only function independent of b which satisfies (2.2).

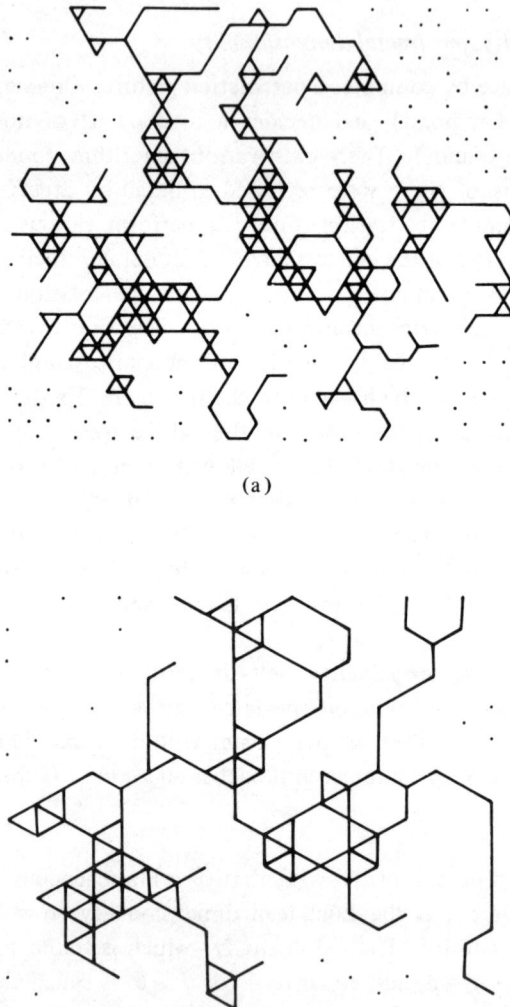

(a)

(b)

Fig. 1. Site percolation on a triangular lattice, at $p_c = \frac{1}{2}$. The sites on the largest connected cluster are emphasized by the connecting bonds. (a) Original simulation. (b) Coarse-grained version, with triangular cells of three sites being occupied by a majority rule.

We note that the relation (2.1) applies only to the *average*. Measurements around individual different central points may result in *fluctuations* around the average, and in variations in the amplitude A. The mean square deviation turns out to behave as (Stauffer, 1980; Kapitulnik, Frid and Deutscher, 1984)

$$\langle M^2 \rangle - \langle M \rangle^2 = CL^{2D} \quad , \tag{2.3}$$

and C is a measure of *lacunarity*, i.e., the size distribution of "lakes" of empty sites within the large cluster (Mandelbrot, 1982). We shall return to this point below.

We should also note that the power law (2.1) is only *asymptotic*, for large L. At realistic values of L this should be modified by *correction terms* (e.g. Margolina *et al.*, 1982)

$$M(L) = AL^D + A_1 L^{D_1} + A_2 L^{D_2} + \dots \quad , \tag{2.4}$$

with $D > D_1 > D_2 > \dots$ It is clearly not easy to extract the correction terms from a direct simulation. Recently, Aharony *et al.* (1985) proposed a novel transfer matrix method, which may make this task easier. In any case, one expects the exponents D_1, D_2, etc. to represent fractal dimensionalities of subgroups of sites on the cluster. For example, the fractal dimensionality of the sites on the cluster which are cut by a random $(d-1)$-dimensional hyperplane should be equal to $(D-1)$, and one expects this value to appear as one of the correction powers. We shall discuss other corrections below.

2.2. Finite clusters at percolation

Consider now a *finite* cluster, containing s sites. Its linear size may be characterized, e.g., by its *radius of gyration*, defined as the root mean square distance between pairs of sites on it (Essam, 1980).

$$R_s^2 = \sum_{i,j} (\mathbf{r}_i - \mathbf{r}_j)^2 / 2s^2 \quad . \tag{2.5}$$

Any other characteristic length, e.g., the size of a cubic box touching the perimeter of the cluster, L_s, scales the same way as R_s. On scales L which are small compared to R_s, the average "mass" of the cluster scales as L^D, as discussed above (the "largest" cluster in Fig. 1 is also finite on some large scales, since the probability of finding an infinite cluster at p_c vanishes). However, this mass should approach the cluster's total mass, s, for $L > R_s$. Dimensionally, this must imply that $s \sim R_s^D$. Since R_s is the only available length, we expect $M_s(L)$ to depend on L only via the ratio L/R_s, hence

$$M_s(L) = R_s^D m(L/R_s) \quad , \tag{2.6}$$

where the scaling function $m(x)$ must approach a finite constant for $x \gg 1$ and

$m(x) \sim x^D$ for $x \ll 1$. This is the first example of a *scaling* argument, which will be used repeatedly below.

Indeed, the relation $s \sim R_s^D$ has been confirmed by many computer simulations. Figure 2 shows the function $s(L_s)$ of such a simulation, on the square lattice, in which individual clusters were grown around a central point, adding sites to the perimeter with probability $p_c = 0.5927$. The error bars indicate the scatter of values of s within "windows" of L_s, which again emphasizes the fact that the power law $s = \bar{A} L_s^D$ applies only to the *average*. The fact that the error bars have a fixed length on the logarithmic scale shows that the second

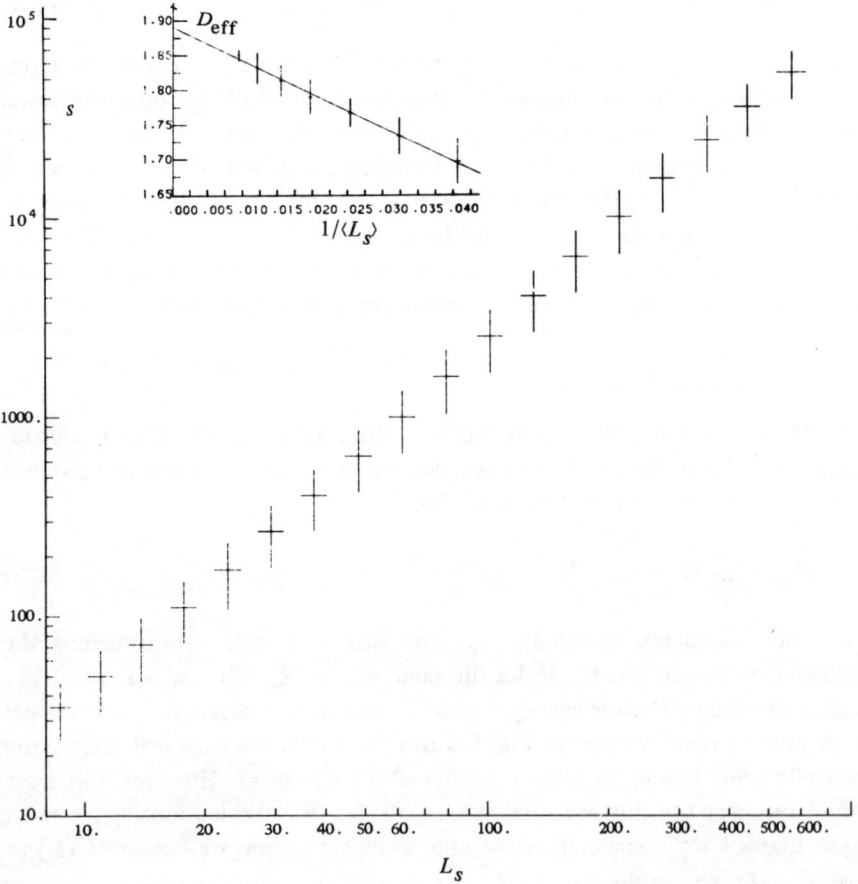

Fig. 2. Dependence of cluster masses s (number of sites) on their linear size L_s, for the square lattice at $p_c = 0.5927$. The "error bars" indicate one standard deviation around the average. The insert shows the extrapolation of local slope $\partial \ln s / \partial \ln L_s$ to $s \to \infty$, where $D \simeq 1.89 \pm 0.01$ (from Grossman and Aharony, 1986).

cumulant of the size distribution, related to the *lacunarity*, behaves as

$$\langle s^2 \rangle - \langle s \rangle^2 \sim L_s^{2D} \quad . \tag{2.7}$$

Similarly, we find that the kth cumulant scales as L_s^{kD} .

Note that the slope in Fig. 2 is not constant. The insert shows the dependence of the local slope, $D_{eff} = d \ln s/d \ln L_s$, on $1/L_s$, and its dependence on L_s indicates the importance of correction terms, as in Eq. (2.4). The asymptotic slope, for $L_s \to \infty$, approaches $D \simeq 1.89 \pm 0.01$. The fact that D_{eff} can be fitted by a straight line in $1/L_s$ indicates that the leading correction D_1 in Eq. (2.4) is close to $D - 1$.

Consider now the *size distribution of finite clusters*. Let n_s be the average number per site of clusters containing s sites. Thus, sn_s is the probability per site to belong to a cluster of s sites. Self-similarity implies that sn_s should have the same dependence on s in the two parts of Fig. 1. However, the sizes of large clusters in Fig. 1b are smaller by a factor b^D, compared to their values in Fig. 1a. Similarly, the total number of sites decreased by a factor b^d. Thus,

$$sn_s = b^{-d}(s/b^D)n_{s/b^D} \quad , \tag{2.8}$$

which immediately implies that

$$n_s \sim s^{-\tau} \quad , \tag{2.9}$$

with

$$\tau = (d + D)/D \quad . \tag{2.10}$$

The fractal dimensionality D thus determines the exponent τ, and through it all the other exponents at p_c .

We note here that the relation (2.10) was derived using the strong assumption that n_s depends *only* upon s. For dimensionalities $d > 6$, this turns out to be wrong: n_s depends there on an additional "dangerous irrelevant" variable, which is related to the density of nodes at which three bonds meet. This yields $\tau = \frac{5}{2}$, $D = 4$ for all $d > 6$. Details of this are presented below, in Sec. 2.6. Relations like (2.10), which are valid only for $d < 6$, are called *"hyperscaling"* relations.

We also note that Eq. (2.8) assumes that the connectivity of clusters does not change under the transformation. As we shall discuss below, in Sec. 3.2, this should be modified. However, the singular part in n_s still obeys a scaling relation like (2.8).

2.3. Finite size scaling

We can next restrict ourselves to a *finite sample*, of linear size L ($L = 27$ in Fig. 1a). The function n_s must also depend on L, and comparison between Figs. 1a and 1b now yields

$$n_s(L) = b^{-d-D} n_{s/b^D}(L/b) \quad . \tag{2.11}$$

Eliminating b, we thus identify

$$n_s(L) = L^{-d-D} n(s/L^D) \quad . \tag{2.12}$$

The *average* cluster size within such a box behaves as

$$\langle s \rangle_L = \sum_{s=1}^{L^D} s^2 n_s(L) \sim L^{2D-d} \quad , \tag{2.13}$$

which is *different* from the "mass" of the *largest* cluster in the box, $s_{max} \sim L^D$. Similarly, the mean square radius of gyration scales as

$$\langle R_s^2 \rangle = \sum_s R_s^2 sn_s / \sum_s sn_s \sim L^{2-(d-D)} \quad , \tag{2.14}$$

and *not* as L^2 (as first noted by Stauffer, 1979).

Equations (2.13) or (2.14) are examples of *finite size scaling*, used widely to identify critical exponents from computer simulations at p_c.

2.4. Crossover away from p_c

We now move away from p_c. If we repeat the coarse graining of Fig. 1 for $p < p_c$, then more and more renormalized sites become empty. Similarly, for $p > p_c$ more and more sites become occupied. The connectedness length ξ, defined as the root mean square distance between occupied sites which belong to the same finite cluster, is now finite, and decreases under repeated iteration.

Every site on the lattice has probability sn_s to belong to a cluster of s sites. If it belongs to such a cluster, it is connected to s other sites. Therefore,

$$\xi^2 = \sum_s R_s^2 s^2 n_s / \sum_s s^2 n_s \quad , \tag{2.15}$$

and n_s must decay quickly for $R_s \gg \xi$, or for $s \gg \xi^D$. In the same spirit as above we may write the scaling form

$$n_s(p) = \xi^{-d-D} \mathscr{N}(s/\xi^D) \quad , \tag{2.16}$$

with $\mathcal{N}(x)$ decaying quickly for $x \gg 1$ and behaving as $x^{-\tau}$ for $x \ll 1$. Thus, ξ serves as an effective cutoff for the size distribution of the finite clusters. Using Eq. (2.16), we may now calculate various averages. For example, the mean cluster size scales as

$$\langle s \rangle = \sum_s s^2 n_s \sim \xi^{2D-d} \quad . \tag{2.17}$$

Similarly, higher moments of the distribution behave as

$$\Gamma_k = \langle s^{k-1} \rangle = \sum_s s^k n_s \sim \xi^{kD-d} \quad . \tag{2.18}$$

The total probability to belong to any finite cluster is $\sum_s s n_s$. Thus, the probability to belong to the "infinite" cluster is

$$P_\infty = p - \sum_s s n_s \quad , \tag{2.19}$$

and the singular part here scales as

$$P_\infty \sim \xi^{D-d} \quad . \tag{2.20}$$

We can now return to the scaling of the "mass" on the "infinite" cluster. For boxes of size $L \ll \xi$ and $p > p_c$ the behavior remains as discussed above, $M(L) \sim L^D$. However, for $L \gg \xi$ the sizes of the finite clusters (and therefore also the sizes of the "holes" in the largest cluster) are cut off at ξ. If we repeat the coarse graining of Fig. 1 l times, so that $b^l = \xi$, then those "holes" will shrink away and the lattice will become full and homogeneous, with density ξ^{D-d} as calculated above. Thus, $M(L)$ should cross over from the fractal behavior, L^D, to the uniform one, $\xi^{D-d} L^d$, via a scaling function of the form

$$M(L) = L^D m(L/\xi) \quad , \tag{2.21}$$

with $m(x) \to$ const for $x \ll 1$ and $m(x) \sim x^{d-D}$ for $x \gg 1$. Indeed, Fig. 3 shows the dependence of $M(L)/L^2$ on L in a simulation of the square lattice, and the crossover at $L \sim \xi$ is very clear.

2.5. Pair connectedness function

The connectedness length ξ is closely related to the pair connectedness function $G(r)$, which is the probability that two sites at a distance r belong to the same cluster. Let $\rho_s(r)$ be the conditional probability that a site at a distance r from

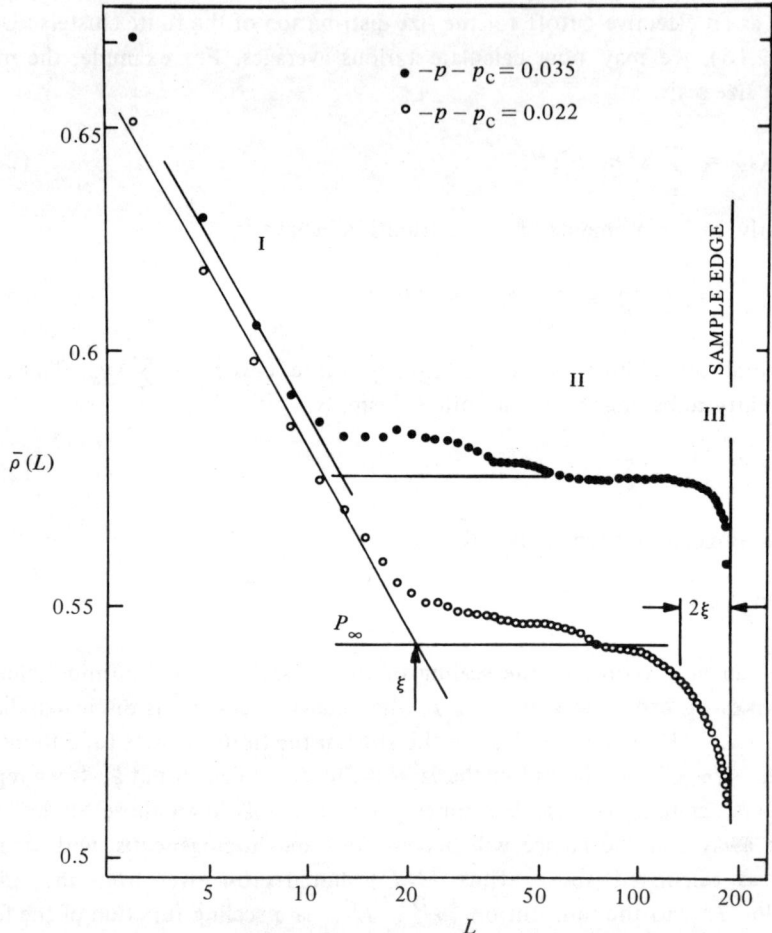

Fig. 3. Density of sites on the largest cluster of a square lattice, at $p - p_c = 0.035$ (full dots) and $p - p_c = 0.022$ (empty dots), within a box of size L around an occupied site. The slope for $L < \xi$ is $D \simeq 1.9$, and the plateau for $L > \xi$ is P_∞ (from Kapitulnik *et al.*, 1983).

the origin belongs to a cluster of s sites, given that the origin belongs to it. We may thus write

$$G(r) = \sum_s s n_s \rho_s(r) + P_\infty \rho_\infty(r) - P_\infty^2 \quad . \tag{2.22}$$

From our earlier discussion, $\rho_s(r)$ decays exponentially for $r > R_s$, and n_s decays exponentially for $R_s > \xi$. For $r < R_s \ll \xi$, all clusters look the same, and

$\rho_s(r) \simeq \rho_\infty(r) \sim r^{D-d}$. At $p = p_c$ this immediately yields

$$G(r) \sim r^{D-d} \sum_{s > r^D} s n_s \sim r^{2(D-d)} \quad . \tag{2.23}$$

In analogy to critical phenomena, this is written as

$$G(r) \sim r^{-(d-2+\eta)} \tag{2.24}$$

and we have identified

$$d - 2 + \eta = 2(d - D) \quad . \tag{2.25}$$

For $p \neq p_c$, Eq. (2.25) may be generalized into the scaling form

$$G(r) = r^{-(d-2+\eta)} g(r/\xi) \quad , \tag{2.26}$$

with an exponential decay of $g(x)$ for $x \gg 1$.

2.6. Behavior at high dimensions

At high dimensions, the probability w that three bonds meet at one site is low. Therefore, the percolating clusters are composed of quasi-one-dimensional branches, which perform random walks. Consider now two connected sites, at a linear distance L from each other, at p_c. They are most probably connected only via a single link, which has L^2 sites (as in a random walk). This link is called the *"backbone"* of the route between the two sites. A fraction w of these sites will have nodes, into "dangling" chains. Within a box of size L, the length of each such chain is also of order L^2, and hence (Aharony, Gefen and Kapitulnik, 1984)

$$M(L) \sim w L^4 \quad . \tag{2.27}$$

Thus, $D = 4$ for high d.

This argument remains self-consistent only if none of the dangling chains meet each other, to form loops. However, the probability of such a meeting is the product of L^2, the typical number of sites on the chain, times $w L^4$, the number of sites on the cluster, times w, the probability of a node, divided by the volume, L^d. The result, $w^2 L^{6-d}$, decays to zero for $d > 6$. This identifies $d_u = 6$ as the *upper critical dimensionality*, above which $D \equiv 4$ (Toulouse, 1974; Kirkpatrick, 1976; Skyes *et al.*, 1976).

For $p > p_c$, the result (2.27) should crossover to $M = P_\infty L^d$, and this can happen only if one uses a scaling function with *two* variables,

$$M(L) = w L^4 m(L/\xi, w^2 L^{6-d}) \quad . \tag{2.28}$$

Although the second argument, $w^2 L^{6-d}$, becomes very small for large L, the only form of $m(x, y)$ which reproduces L^d for all d is $m(x, y) \sim x^2/y$, which is singular for $y \to 0$. This is the reason why the "dangerous irrelevant" variable $y = w^2 L^{6-d}$ cannot be ignored. Substituting $m \sim x^2/y$ in (2.28) we can now identify the dependence of P_∞ on ξ, $P_\infty \sim 1/(w\xi^2)$.

Returning now to the relation (2.19), the singular part in P_∞ should result from $\Sigma s n_s$, with the sum bounded by $s(\xi) \sim w\xi^4$. This can be achieved only if

$$n_s \sim w^{-1/2} s^{-5/2} \quad , \tag{2.29}$$

i.e., $\tau = \frac{5}{2}$ for $d > 6$. The result (2.29) can also be derived using space rescaling, as in Eq. (2.8). Since now $s \sim w R_s^4$, a coarse graining by a factor b turns (s/w) into $b^{-4}(s/w)$. Similarly, n_s should be rescaled by b^{-d-4}. Since w^2 scales into $w^2 b^{6-d}$, (2.29) is the only form which obeys the relation $n_{s/w}(w^2) = b^{-d-4} n_{b^{-4}s/w}(w^2 b^{6-d})$.

All of these arguments break down for $d < 6$, when the node density w effectively increases with the length scale. As we shall see below, w eventually rescales into a finite fixed point value, and one recovers the hyperscaling relations, e.g., (2.10) or (2.25).

2.7. Backbone, graph distance, singly connected bonds and other geometrical properties

So far we characterized the percolating clusters only by their total "mass" s and linear size, characterized by R_s. For our physical applications we find many additional geometrical properties quite useful. Given two points at two ends of the cluster, there are many self-avoiding routes on the cluster which connect them. However, these routes do not cover the whole cluster: any bond which is connected to the cluster only at one of its ends does not carry current and therefore is not a part of such a self-avoiding route. Eliminating all these *"dangling"* bonds, we are left with the *"backbone"* of the cluster (which depends on the two chosen end points). It is easy to convince oneself that at p_c, the backbone is also self-similar, and thus may be characterized by a fractal dimensionality, D_B, which describes the scaling of the backbone "mass" with the distance between the end points,

$$M_B(L) \sim L^{D_B} \quad . \tag{2.30}$$

Clearly, $D_B < D$. For dimensions $d > 6$, the above discussion shows that the backbone is a random walking quasi-one-dimensional link, with $D_B = 2$. At $d = 2$, current estimates yield $D_B \simeq 1.6 - 1.7$ (Hermann and Stanley, 1984; Hong and Stanley, 1984). Thus, most of the mass of the cluster belongs to the dangling bonds.

At dimensionalities $d < 6$, the backbone contains *loops,* so that there exist several routes between the two end points. The shortest of these (measured by counting bonds or sites on it) is called the *graphical,* or *topological,* or *chemical* distance L_{chem} (e.g., Alexandrowicz, 1980). This distance is relevant, e.g., for the propagation of a disease in a dilute forest. Self similarity again implies a power law at p_c,

$$L_{chem} \sim L^{\tilde{\zeta}_{chem}} \quad . \tag{2.31}$$

For $d > 6$, $\tilde{\zeta}_{chem} = D_B = 2$. Generally, $1 \le \tilde{\zeta}_{chem} \le D_B$. At $d = 2$, $\tilde{\zeta}_{chem} \simeq 1.15$ (e.g., Havlin and Nossal, 1984).

As d decreases, there appear more and more loops, or "blobs," of bonds. In an extreme picture, the whole backbone will involve loops within loops. In practice, this turns out to be too extreme: the backbone contains bonds which are common to *all* the routes between the two end points. Disconnecting any of these bonds will disconnect the two end points. Denoting the number of these *"singly connected"* bonds by L_{sc}, we again expect at p_c (Pike and Stanley, 1981)

$$L_{sc} \sim L^{\tilde{\zeta}_{sc}} \quad . \tag{2.32}$$

with $\tilde{\zeta}_{sc} \le \tilde{\zeta}_{chem} \le D_B$ (equalities hold at $d > 6$).

As discussed above, all of these relations should apply only for $L < \xi$ on the "infinite" cluster, or for $L < R_s < \xi$ on a cluster of s sites (or bonds). In particular, one expects that the infinite cluster becomes *homogeneous* for $L > \xi$, when M_B should cross over to $P_B L^d$, with $P_B \sim \xi^{D_B - d}$,

$$L_{chem} \sim \xi^{\tilde{\zeta}_{chem} - 1} L \tag{2.33}$$

and

$$L_{sc} \sim \xi^{\tilde{\zeta}_{sc}} \quad . \tag{2.34}$$

Following the pioneering works of Skal and Shklovskii (1975) and de Gennes (1976), a commonly accepted model of the backbone uses the *node-link-blob picture* (Stanley, 1977; Coniglio, 1981): the backbone is a homogeneous superlattice of nodes, separated by a linear distance ξ and two neighboring nodes are connected via a single link which has both singly connected bonds and blobs of multiconnected ones.

Coniglio (1981, 1982) made the very important observation that in this limit, $L_{sc} \sim (p - p_c)^{-1}$. Combined with (2.34), this yields the power law divergence

$$\xi \sim (p - p_c)^{-\nu} \quad , \tag{2.35}$$

with

$$\tilde{\zeta}_{sc} = 1/\nu \quad . \tag{2.36}$$

We shall return to this relation below, in Sec. 3.2.

In addition to the total number of singly connected bonds, L_{sc}, one may also consider the distribution of the lengths of their individual segments between neighboring "blobs" (Herrmann and Stanley, 1984). The number of these segments, which is equal to the number of "blobs," also scales as $L_{sc} \sim L^{\tilde{\zeta}_{sc}}$ (Grossman and Aharony, 1986), showing that the average length of these segments is finite. On the other hand, the sizes of the blobs have a power law distribution, similar to n_s (Hermann and Stanley, 1984).

We may also consider pairs of *"doubly connected" bonds,* such that disconnecting the two bonds in a pair disconnects the two end points. The number of these pairs, L_2, also obeys a power law scaling, and one can show that (Grossman and Aharony, 1986)

$$L_2 \sim L^{2\tilde{\zeta}_{sc}} \sim L^{2/\nu} \quad . \tag{2.37}$$

Another property which depends on the distance between two end points concerns the *electrical resistance, R,* given that each bond has a unit resistance. We have recently considered nonlinear resistors, in which the voltage V scales as a power of the current I, $|V| = r |I|^{\alpha}$, and again confirmed a power law,

$$R \sim L^{\tilde{\zeta}(\alpha)} \tag{2.38}$$

(Blumenfeld and Aharony, 1985). This resistance approaches L_{chem} when $\alpha \to 0^+$ and L_{sc} when $\alpha \to \infty$. We shall return to the exponent $\tilde{\zeta}_R = \tilde{\zeta}(1)$ below, in Sec. 5.

All the properties discussed so far were defined via two end points. The resulting quoted power laws were again meant to describe the *averages* over many pairs of points, with a fixed distance L. Alternatively, one could consider finite clusters, and study these and other properties as a function of their size, s. The results are again power laws, in which L is replaced by $s^{1/D}$ (Grossman and Aharony, 1986). Other properties are defined for the cluster as a whole. An example which drew some attention concerns the external perimeter or the "hull" of the cluster, which is also found to scale as a power of s, with an exponent 0.93 in $d = 2$ (Voss, 1984).

2.8. Fractal models

In what follows we aim to solve various physical problems on the dilute system. Having realized the self-similar structure of the percolation clusters, it has been very useful to construct *non-random fractal models,* which imitate many of the geometrical properties of these clusters. The first example, aimed to imitate the *backbone,* concerns d-dimensional Sierpinski gaskets (Gefen *et al.,* 1981). Figure 4 shows four stages of the recursive construction of the two-dimensional case: at each iteration, the length scale is divided by $b = 2$, and each triangle is replaced by three smaller ones (the central small triangle is cut out). As explained above, the fractal dimensionality is given by $3 = 2^D$, i.e., $D = \ln 3 / \ln 2 \simeq 1.58$. For the d-dimensional generalization (using hyper-tetrahedra),

$$D = \ln(d + 1)/\ln 2 \quad , \tag{2.39}$$

and for $d < 4$ this gives a reasonable estimate of the backbone fractal dimensionality D_B. By construction, the graphical distance is always a straight line, i.e., $\tilde{\zeta}_{chem} = 1$. The structure contains no singly connected bonds, hence $\tilde{\zeta}_{sc} = 0$. Although this latter property is not representative of the real backbones, the gaskets have been very useful in obtaining a feeling for the effects of the loop-within-loop geometry on many physical problems, ranging from magnetic properties (Gefen *et al.,* 1984), conductivity (Gefen *et al.,* 1981) and quantum wave functions (Domany *et al.,* 1983).

Fig. 4. Four stages of iteration of the Sierpinski gasket.

A more flexible model for the two-dimensional full cluster was constructed by Mandelbrot and Given (1984), based on branching Koch curves. The model, shown in Fig. 5, has $D = \ln 8/\ln 3 \simeq 1.89$, $D_B = \ln 6/\ln 3 \simeq 1.63$, $\tilde{\zeta}_{sc} = \ln 2/\ln 3 \simeq 0.63$, $\tilde{\zeta}_{chem} = 1$ and $\tilde{\zeta}_R = \ln(11/4)/\ln 3 \simeq 0.92$. This model was further generalized by de Arcangelis *et al.* (1985). Varying the dependence of the lengths of the various segments in the branching Koch curve on the rescale factor b with d, one may imitate the values of e.g., ζ_{chem} and $\tilde{\zeta}_{sc}$ or $\tilde{\zeta}_{sc}$ and D_B. One may then solve for other properties, and modify the model as necessary.

3. Phase Transitions

3.1. *Dilute magnets*

The above discussion has already indicated many analogies with thermal phase transitions: the probability of belonging to the "infinite" cluster P_∞ is zero for $p < p_c$, and grows for $p > p_c$, similarly to the temperature dependence of the order parameter in a magnetic phase transition. In fact, combining (2.20) with (2.35) we may write

$$P_\infty \approx B|t|^\beta \quad , \quad t \to 0^- \ , \tag{3.1}$$

with $t = (p_c - p)/p_c$ and (for $d < 6$)

$$\beta = (d - D)\nu \ . \tag{3.2}$$

Similarly, the moments Γ_k (see (2.18)) will diverge as

$$\Gamma_k \approx C_k^{\pm}|t|^{-\gamma_k}, \quad t \to 0 \ , \tag{3.3}$$

with $+(-)$ denoting $t > 0 \ (< 0)$ and

$$\gamma_k = (kD - d)\nu \ . \tag{3.4}$$

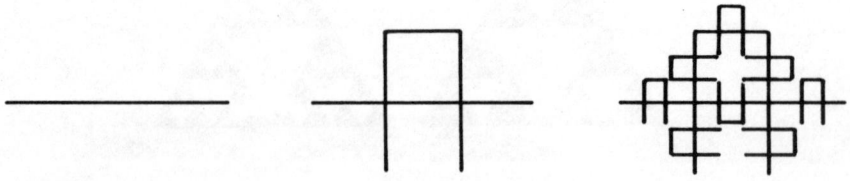

Fig. 5. Two stages of iteration of the Mandelbrot-Given model.

To make this analogy more explicit, consider the dilute Ising model, with Hamiltonian

$$\mathcal{H} = - \sum_{\langle ij \rangle} J_{ij} S_i S_j \quad , \tag{3.5}$$

where $S_i = \pm 1$ and where J_{ij} are quenched given nearest neighbor exchange coefficients, which take the values J (with probability p) or zero (with probability $1 - p$). At zero temperature, the ground state will have all the spins which belong to the same connected cluster parallel to each other. The spin correlation function will thus be

$$\langle S_i S_j \rangle = \begin{cases} 1 & i, j \text{ belong to same cluster,} \\ 0 & \text{otherwise.} \end{cases} \tag{3.6}$$

Averaging over all pairs of spins, this becomes exactly $G(r)$, discussed in Sec. 2.5. Summing over all pairs within a cluster of s sites, $\sum_{i,j} \langle S_i S_j \rangle = s^2$, and thus the average magnetic susceptibility of the finite clusters behaves as $\sum_s n_s s^2$, i.e., as Γ_2. Below p_c, the average magnetization is zero. Above p_c, all the finite clusters do not contribute to the average magnetization, which then becomes equal to the density of the infinite cluster, i.e., to P_∞.

To draw further analogies, it is convenient to use the Laplace transform of $n_s(p)$,

$$F(p, h) = \sum_s n_s(p) e^{-sh} \quad . \tag{3.7}$$

$F(p, 0)$ is immediately identified as the mean number (per site) of finite clusters. All the moments discussed above are now obtained as derivatives of F:

$$P_\infty = p - \sum_s s n_s = p + \left. \frac{\partial F}{\partial h} \right|_{h=0} \quad , \tag{3.8}$$

$$\Gamma_k = \langle s^{k-1} \rangle = \sum_s s^k n_s = \left. -\left(\frac{\partial}{\partial h} \right)^k F \right|_{h=0} \quad . \tag{3.9}$$

The analogy of P_∞ to the order parameter thus identifies F as the "free energy,"

h as the ordering field, Γ_2 as the susceptibility, G as the correlation function, etc. A geometric interpretation of h arises if h represents the probability that an occupied site (or bond) is connected to a single "ghost site," which is not on the lattice.

All the scaling relations discussed above may now be rewritten in terms of the singular part of the function F. In particular, Eqs. (2.8) or (2.16) now become

$$F_{\mathrm{sing}}(\xi, h) = b^{-d}\,\overline{\mathscr{F}}(\xi/b, b^D h) \quad . \tag{3.10}$$

Substituting (2.35), or

$$\xi = \xi_0^{\pm}\,|t|^{-\nu} \quad , \tag{3.11}$$

this becomes

$$F_{\mathrm{sing}}(p, h) \approx |t|^{2-\alpha}\,\mathscr{F}(h\,|t|^{-\Delta}) \quad , \tag{3.12}$$

with (e.g., Nakanishi and Stanley, 1978; Aharony, 1980)

$$\Delta = D\nu \quad , \tag{3.13}$$

$$2 - \alpha = d\nu \quad . \tag{3.14}$$

Equation (3.12) is exactly the same as that arising in critical phenomena, and all the usual tools used there may be readily transferred here. There exist two main routes to do this: First, one may use renormalization group ideas directly on the geometrical problem. An introduction to this approach is given in Sec. 3.2. Alternatively, one may use the dilute magnet to map the problem directly on a spin Hamiltonian, and then proceed using statistical mechanics. This is described in Sec. 3.3.

3.2. Real space renormalization group

A simple version of the *real space renormalization group* was given in Fig. 1. In the transformation from Fig. 1a to Fig. 1b we replaced each cell by an occupied site if a majority of its three sites were occupied. The probability of having an occupied renormalized site is thus (Reynolds *et al.*, 1980)

$$p' = p'(p) = p^3 + 3p^2(1-p) \quad . \tag{3.15}$$

Assuming that the new Fig. 1b can be fully described by the parameter p', we can iterate this transformation many times. There are three values of p which represent *fixed point* of the transformation (3.15), i.e., they remain invariant under iteration: $p^* = 0$, $\frac{1}{2}$, 1. All the points with $p < \frac{1}{2}$ "flow" towards zero, and all the points with $p > \frac{1}{2}$ "flow" towards unity. This agrees with our qualitative expectations, as discussed in Sec. 2.4.

We now turn to the connectedness length, ξ. If ξ is fully determined by p, then we expect that

$$\xi(p') = \xi(p)/b \quad , \tag{3.16}$$

due to the change of the length unit by b. At a fixed point, $p' = p$ and thus ξ must be 0 or ∞. Indeed, $\xi = 0$ at the trivial fixed points $p^* = 0$, 1. The additional point, $p^* = \frac{1}{2}$, is now identified as the percolation threshold, with $\xi = \infty$. Although the recursion relation (3.15) is approximate, the result $p_c = \frac{1}{2}$ happens to be exact.

Near p_c, we may linearize (3.15), and write

$$t' \doteq \Lambda_t t \quad , \tag{3.17}$$

with $t = (p_c - p)/p_c$ and $\Lambda_t = (\partial p'/\partial p)_{p^*}$. Thus $\xi(\Lambda_t t) = \xi(t)/b$, and $\xi(t)$ must have the power law form (2.35), with $1/\nu = \ln \Lambda_t /\ln b$.

In the example (3.15), $\Lambda_t = \frac{3}{2}$ and $\nu = \ln 3^{1/2} /\ln(\frac{3}{2}) \simeq 1.355$, which is an excellent approximation to the exact two-dimensional value $\nu = \frac{4}{3}$ (found by den Nijs, 1979).

The recursion relation (3.15) is approximate, since it does not always preserve the connectivity of clusters. For example, Fig. 6a shows two disconnected clusters which are combined into one cluster in the coarse grained picture. To preserve the identity of clusters, we should, e.g., introduce a new bond concentration variable, which will indicate the probability that two neighboring renormalized sites are connected (see e.g., Nakanishi and Reynolds, 1979). Similarly, Fig. 6b shows a portion of a single connected cluster, which splits into two disconnected clusters in the renormalized picture. Examples of additional parameters include next nearest neighbor connectivity, the probability of three bond nodes (as discussed in Sec. 2.6), etc. In general, one thus "flows" into a multidimensional parameter space $\{p_1, p_2, p_3, \ldots\}$. It is usually impossible to identify exactly a finite number of p_i's under which the renormalization group recursion relations become closed. All real space renormalization group calculations (except in $d = 1$) must therefore use truncations, and involve approximations. For a review, see, e.g., Stanley *et al.* (1982).

Although one cannot carry out an exact real space calculation, one may still make several general statements. Since $\xi(p_c) = \infty$, ξ will remain infinite

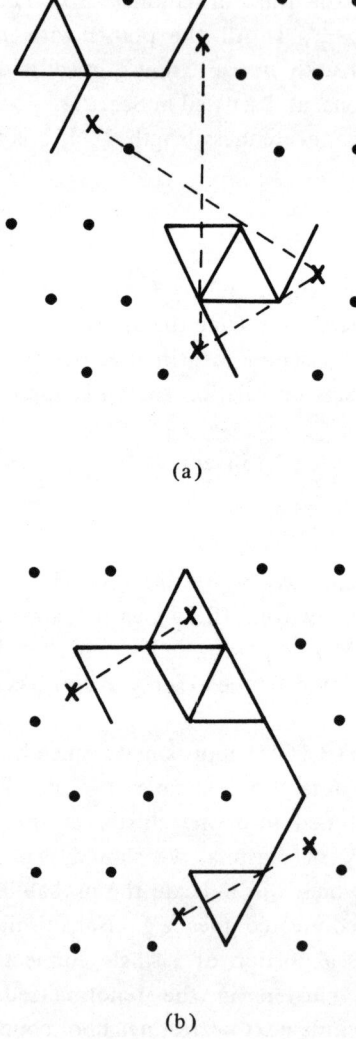

(a)

(b)

Fig. 6. Examples of renormalization group configurations. (a) Two disconnected clusters which become connected. (b) A connected cluster which splits into two disconnected clusters. Full lines indicate connected "old" sites. Broken lines show connected "new" sites.

on the trajectory into which the point $p = p_c$ will "flow." Experience with critical phenomena then leads to the expectation that this trajectory approaches a *fixed point* in the multidimensional space, $p_1^*, p_2^*, \ldots$ In fact, one expects all the points on a critical hyper-surface to flow to this fixed point, and this is confirmed by all the approximate real space calculations.

Near the fixed point, the renormalization group transformation may be linearized,

$$p'_i - p_i^* \approx \sum_j L_{ij}(p_j - p_j^*) \quad . \tag{3.18}$$

Diagonalizing the matrix L_{ij} yields *linear scaling fields*, $\{\mu_i\}$, which renormalize via

$$\mu'_i = \Lambda_i \mu_i + O\{\mu_i^2\} \quad . \tag{3.19}$$

Since $\Lambda_i(b^2) = [\Lambda_i(b)]^2$, we have $\Lambda_i = b^{\lambda_i}$. The largest eigenvalue is now identified as $\Lambda_1 = \Lambda_t = b^{1/\nu}$, and the other eigenvalues usually obey $\Lambda_i < 1$, i.e., $\lambda_i < 0$. The corresponding parameters μ_i are "irrelevant," since they decay to zero after many iterations.

Similar to the concentration-like variables, p_i, the renormalization group transformation will also generate additional "ghost" fields, h_i. The largest eigenvalue associated with these is identified as $\Lambda_h = b^D$, and the others are again "irrelevant." Our Eq. (3.10) must now be generalized,

$$F_{\text{sing}}\{\mu_i, h_i\} = b^{-d} F_{\text{sing}}\{\mu'_i, h'_i\} \quad . \tag{3.20}$$

A detailed discussion of this equation will be given below, in Sec. 3.4.

Before concluding this section, we return briefly to Eq. (2.36), which relates the "thermal" exponent $1/\nu$ to the geometrical one, $\tilde{\zeta}_{sc}$. Following Coniglio (1981, 1982), consider a bond percolation cluster, at p_c. Consider now a "renormalized" bond, connecting two points at a linear distance b apart. If we eliminate a fraction q of the bonds, then the probability that the two end points become disconnected is $q' = L_{sc} q + O(q^2)$, where $L_{sc} = b^{\tilde{\zeta}_{sc}}$ is the number of singly connected bonds on the backbone connecting the two points. Since $p = p_c(1-q), q = (p_c - p)/p_c = t$, and we find $t' = b^{\tilde{\zeta}_{sc}} t$. Comparison with (3.17) thus identifies $\Lambda_t = b^{1/\nu} = b^{\tilde{\zeta}_{sc}}$, and proves Eq. (2.36).

As reviewed by Stanley *et al.* (1982), many real space renormalization group calculations have been performed on the problem, and the resulting values of D and ν improve as the number of parameters used is increased. More recently, the method of transfer matrices on strips has also proved rather accurate (Vannimenus and Nadal, 1984).

3.3. Potts model, field theory and universality

There are many ways to show the mapping of the percolation problem to the $q \to 1$ limit of the q-state Potts model (Kasteleyn and Fortuin, 1969). We choose here to start from the dilute magnetic Hamiltonian, Eq. (3.5). The average free energy density is

$$-\beta F = \frac{1}{N} \left[\ln Z \{J_{ij}\} \right]_{av} \quad , \tag{3.21}$$

where $\beta = 1/k_B T$,

$$Z\{J_{ij}\} = \text{Tr} \exp \left(\beta \sum_{\langle ij \rangle} J_{ij} S_i S_j \right) \quad , \tag{3.22}$$

and [ ]$_{av}$ denotes an average over the quenched distribution of J_{ij}'s. We now use the "replica trick," replacing $\ln Z$ by $\lim_{n \to 0} [(Z^n - 1)/n]$. The result is

$$[Z^n]_{av} = \text{Tr} \prod_{\langle ij \rangle} \left[p \exp \left(\beta J \sum_\alpha (S_i^\alpha S_j^\alpha - 1) \right) + 1 - p \right], \tag{3.23}$$

where the trace is now over the n replicated spin ($S_i^1, \ldots, S_i^n$), and where we shifted $\mathcal{H}$ by the constant $\sum_{\langle ij \rangle} J_{ij}$. We now have $[Z^n]_{av} = \prod_{\langle ij \rangle} \exp(h_{ij})$, with

$$\exp(h_{ij}) = p \exp \left[K \sum_\alpha (S_i^\alpha S_j^\alpha - 1) \right] + 1 - p \quad , \tag{3.24}$$

where $K = \beta J$. In the limit $T \to 0$, $K \to \infty$, $\exp(h_{ij}) = 1$ when $S_i^\alpha \equiv S_j^\alpha$ for all α, and $\exp(h_{ij}) = 1 - p$, otherwise. Thus

$$h_{ij} \xrightarrow[K \to \infty]{} -\ln(1-p) \left(\prod_\alpha \delta_{S_i^\alpha, S_j^\alpha} - 1 \right) \quad . \tag{3.25}$$

This is exactly the Hamiltonian of the q-state Potts model, with $q = 2^n$ states and with coupling constant $-\ln(1-p)$. The free energy of the dilute problem, at $T = 0$, may thus be obtained by calculating the partition function of this Potts model, and then taking the limit $n \to 0$, or $q \to 1$.

The mapping onto the Potts model has been very useful for obtaining exact values for the critical exponents at $d = 2$, e.g., $\nu = \frac{4}{3}$, $\beta = \frac{5}{36}$, $\gamma = \gamma_2 = \frac{43}{18}$, $D = \frac{91}{48}$, etc. (den Nijs, 1979; Nienhuis, 1982). It has also been crucial in generating a field theory (e.g., Harris *et al.*, 1975; Amit, 1976). One version of the Potts model field theory is written using the Hamiltonian (Priest and Lubensky, 1976; Pytte, 1979; Aharony, 1980)

$$\mathcal{H} = -\frac{1}{4}\int(r_0 + k^2)\sum_{i=1}^{q} Q_{ii}(\mathbf{k})Q_{ii}(-\mathbf{k})$$

$$+ w\iiint\sum_{i} Q_{ii}Q_{ii}Q_{ii} + \ldots, \tag{3.26}$$

where Q_{ii} is a diagonal traceless $q \times q$ tensor and $\int$ denotes integration over $\mathbf{k}$ in the first Brillouin zone. One can now apply all the standard momentum space renormalization group methods, in which the degrees of freedom $Q_{ii}(\mathbf{k})$ with $\mathbf{k}$ near the edge of the Brillouin zone are gradually traced out (Ma, 1976). The renormalization group recursion relation for w now takes the form (setting $q = 1$)

$$\frac{dw}{dl} = \frac{\epsilon}{2}w(l) - \frac{A}{2}w(l)^3 + \ldots, \tag{3.27}$$

where $\epsilon = 6 - d$, and A is of order unity. The solution for small ϵ and w is

$$w(l)^2 = \frac{w(0)^2 e^{\epsilon l}}{1 + \left[\dfrac{w(0)}{w^*}\right]^2 (e^{\epsilon l} - 1)} \tag{3.28}$$

with $(w^*)^2 = \epsilon/A$.

For $d > 6$, $w(l)$ iterates towards zero, and $w(l)^2 \simeq w(0)^2 b^\epsilon$ is recovered, which is the same as the rescaling of w in Sec. 2.6. Indeed, one can identify w with the probability of finding three-bond vertices. The second term in Eq. (3.27) is the result of combining three vertices into a single "renormalized" vertex (Fig. 7), which lowers the effective density of these vertices. For $d > 6$, this term stops the growth of $w(l)$, and generates a "flow" to the fixed point w^*.

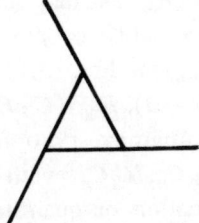

Fig. 7. The renormalization of three vertices into a single new vertex.

The parameter r_0 in Eq. (3.26) is linear in the initial concentration of bonds, p (through the Potts coupling constant $-\ln(1-p)$). The solution of the appropriate recursion relation now yields

$$t(l) = te^{2l}/[1 + (w/w^*)^2 (e^{\epsilon l} - 1)]^{5/21} \quad , \tag{3.29}$$

with $t = r_0 + Bw^2$. For large l this becomes $t(l) \sim te^{l/\nu}$, with

$$2\nu = 1 + \frac{5}{42} \epsilon + \frac{589}{18522} \epsilon^2 + \ldots \tag{3.30}$$

Similar expansions may be obtained for all the other exponents. These now exist up to order ϵ^3. Together with information on the behavior of high order terms, these asymptotic expansions have been extrapolated down to finite values of ϵ, and the results agree remarkably well with those obtained from other sources (de Alcantara Bonfim *et al.*, 1980, 1981). Values of the exponents, collected from various sources, are listed in Table. 1.

At $d = 6$, $\epsilon = 0$, Eq. (3.29) reduces to $t(l) \sim te^{2l} l^{-5/21}$, and hence to $\xi \sim |p - p_c|^{-1/2} |\ln|p - p_c||^{5/42}$. Similar logarithmic corrections arise in all other quantities (Essam *et al.*, 1978; Aharony, 1980; Aharony, Gefen and Kapitulnik, 1984).

In addition to the exponents, the renormalization group and related field theoretical techniques may be used to derive the *equation of state* of the Potts model, which gives the dependence of the order parameter P_∞ on t and h. In the close vicinity of the fixed point one indeed recovers the scaling form (Aharony, 1980)

$$h/P_\infty^\delta = f(t/P_\infty^{1/\beta}) \quad , \tag{3.31}$$

with $\delta = 1 + \gamma/\beta$, which is equivalent to Eq. (3.12) (or to its field derivative). Since the function f is calculated in the vicinity of the fixed point, it turns out to be *universal*, i.e., independent of the initial values of r_0 and w. These enter only into the scales of h and P_∞, and may be removed by the normalizations $f(0) = 1$, $f(-1) = 0$. The universality of $f(x)$ now yields many universal combinations of amplitudes, e.g., (see Eqs. (3.1) and (3.3)) C_k^+/C_k^-, A_F^+/A_F^- (where $F_{\text{sing}} \approx A_F^\pm |t|^{2-\alpha}$ at $h = 0$), $R_\chi \equiv C_2^+ E^{-\delta} B^{\delta-1}$ (where $P_\infty = Eh^{1/\delta}$ at $p = p_c$), $R_C \equiv A_F^+ B^{-2} C_2^+$ (Aharony, 1980 and references therein), and — more recently — ratios like $C_k C_m/C_l C_n$, with $k + m = l + n$ (Adler *et al.*, 1986). In general, any combination of quantities which behaves as $|t|^0$ (a constant) is independent of the two non-universal scale factors, and is therefore universal.

Table 1. Estimates of exponents from various sources.

d	1	2	3	4	5	>6
D	1	$\frac{91}{48}$	2.5	3.2	3.5	4
$\tau=(d+D)/D$	2	$\frac{187}{91}$	2.2	2.25	2.43	$\frac{5}{2}$
D_B	1	1.6-1.7	1.8	2.1	2.3	2
ν	1	$\frac{4}{3}$	0.88	0.64	0.51	$\frac{1}{2}$
$\tilde{\zeta}_{chem}$	1	1.15	1.36	1.7	2	2
$\tilde{\zeta}_R$	1	0.97	1.2	1.6	1.8	2
$\mu=(d-2+\tilde{\zeta}_R)\nu$		1.3	2.0	2.4	2.8	3
s		0.97	0.75-0.85	0.4	0.15	0
β	0	$\frac{5}{36}$	0.44	0.67	0.83	1
γ	1	$\frac{43}{18}$	1.8	1.44	1.2	1
ω		0.9-1	1-1.8	1.3	0.8	$(d-6)$

As with the equation of state, one may also derive a scaling form for the Fourier transform of $G(r)$,

$$\hat{G}(k, t, P_\infty) = |t|^{-\gamma} Z(t\, P_\infty^{-1/\beta}, \; k|t|^{-\nu}) \qquad . \tag{3.32}$$

The normalized $Z(x, y)$ is again universal, and it allows the derivation of universal ratios like ξ_0^+/ξ_0^-, $(R^+)^d \equiv A_F^+(\xi_0^+)^d$, etc. (Aharony, 1980).

Since all of these ratios are determined by the fixed point, which is independent of the detailed initial problem, they should depend only on d, and not on the lattice structure, the type of dilution (bond or site), on the range of correlations among occupied sites or bonds (as long as it is finite), etc.

3.4. Corrections to scaling

Returning to Eq. (3.29), for $d < 6$, we see that if w is initially at its fixed point value then $t(l) = te^{l/\nu}$ exactly, and we conclude that $\xi \sim |t|^{-\nu}$. However, if $w \neq w^*$ then the denominator in (3.29) generates a correction term, which to this order in ϵ has the form

$$\xi = \xi_0^+\, t^{-\nu}(1 + a[1 - (w/w^*)^2]\, t^{\epsilon/2})^{5/42}$$

$$\simeq \xi_0^+\, t^{-\nu}(1 + \tfrac{5}{42} a[1 - (w/w^*)^2]\, t^{\epsilon/2}) \qquad . \tag{3.33}$$

The correction term, $t^{\epsilon/2}$, becomes very small for $t \to 0$. However, it remains important for finite values of t, whenever $[1 - (w/w^*)^2]\, t^{\epsilon/2}$ is not too small. Similar corrections arise for all the quantities, and one may write

$$P_\infty = B|t|^\beta (1 + a_p |t|^{\omega\nu}) \; ,$$

$$F_{\text{sing}} = A_F^\pm |t|^{2-\alpha} (1 + a_F^\pm |t|^{\omega\nu}) \qquad ,$$

$$\Gamma_k = C_k^\pm |t|^{-\gamma_k} (1 + a_k^\pm |t|^{\omega\nu}) \quad , \tag{3.34}$$

etc. Since all the correction amplitudes contain $[1 - (w/w^*)^2]$ and a universal coefficient, the ratios $a_p/a_F^\pm$, $a_\xi^\pm/a_k^\pm$, etc. are also all universal (Aharony, 1980).

The exponent ω in (3.34), which is equal to ϵ to leading order in ϵ, is identified as the exponent describing the rescaling of $[w^2 - (w^*)^2]$ near the fixed point,

$$[w^2 - (w^*)^2]' \simeq b^{-\omega}[w^2 - (w^*)^2] \qquad . \tag{3.35}$$

A similar correction term appears in F_{sing} at $p = p_c$,

$$F_{\text{sing}} \simeq \tilde{A} \, h^{(2-\alpha)/\Delta} \, (1 + \tilde{a}_F \, h^{\omega \nu/\Delta}) \quad . \tag{3.36}$$

Using the inverse Laplace transform, this yields

$$n_s \sim s^{-\tau} (1 + a_n s^{-\omega/D}) \quad . \tag{3.37}$$

Similarly, the correlation function at p_c will have the form

$$G(r) \sim r^{2(D-d)} (1 + a_G \, r^{-\omega}) \quad , \tag{3.38}$$

which may be translated into Eq. (2.4), with (Aharony *et al.*, 1985)

$$D_1 = D - \omega \quad . \tag{3.39}$$

In dimensions $d = 2$ and $d = 3$, existing estimates yield $\omega = 1$ (Houghton *et al.*, 1978; Margolina *et al.*, 1984), which yield $D_1 \simeq D - 1$, probably interpretable as the fractal dimension of a random hyperplanar cut.

In addition to w there may exist many other irrelevant variables, as discussed following Eq. (3.19). It is only near $d = 6$ that w turns out to overcome all the others, with the four-bond vertex having an exponent ω_4 of order $2 + O(\epsilon)$. In general, Eq. (3.20) may be rewritten as

$$F_{\text{sing}} = |\mu_1|^{2-\alpha} \mathscr{F}(h_1 |\mu_1|^{-\Delta} , \mu_i |\mu_1|^{\omega_i \nu} , h_i |\mu_1|^{\theta_i \nu}) \quad , \tag{3.40}$$

with $\omega_i = -\lambda_i > 0$ (and similarly $\theta_i > 0$) for $i > 2$. The dimensionalities D_i in Eq. (2.4) are thus identified as $D - \omega_i$ or $D - \theta_i$. Their full geometrical interpretation remains to be studied.

In addition to the correction terms which arise from irrelevant variables, there also exist corrections which arise from the *nonlinear terms* in Eq. (3.19) and from the mapping of the original variables t and h onto μ_1 and h_1 (Aharony and Fisher, 1983). Equation (3.19) may be rewritten as

$$g_i' = \Lambda_i g_i \quad , \tag{3.41}$$

where the *nonlinear scaling fields* are analytic functions of the original variables, e.g.,

$$g_1 = c_1 t + c_2 h + O(th, t^2, h^2) \quad ,$$

$$h_1 = c_3 h + O(h^2, th) \quad . \tag{3.42}$$

In terms of the g_i's F_{sing} is exactly a homogeneous function. Taking derivatives with respect to h, or expanding in the nonlinear terms, now implies correction terms. In particular, if $c_2 \neq 0$ this generates a correction term in Eq. (2.4) with $D_x = 1/\nu$, i.e., the fractal dimensionality of the singly connected bonds! The terms of order t^2 in g_1 will generate *analytic* corrections, with $\omega\nu$ replaced by 1, and thus with $D - \omega$ replaced by $D - 1/\nu$ (Aharony, Gefen, Kapitulnik and Murat, 1985).

3.5. Low concentration series

The cluster numbers $n_s(p)$ are simple polynomials in p. For bond percolation on a hypercubic d-dimensional lattice one has (e.g., Aharony and Binder, 1980 and references therein)

$$n_1 = dp(1-p)^{4d-2} \quad ,$$

$$n_2 = d(2d-1)p^2(1-p)^{6d-4} \quad , \tag{3.43}$$

etc. Similar expressions can be constructed for site percolation (e.g., Gaunt and Ruskin, 1978). Having identified n_s for a finite range of values of s (e.g., up to $s = 11$), one can substitute the results in sums like (3.9), and obtain expansions of the form

$$\Gamma_k = \sum_i a_i^k(d)p^i \quad , \tag{3.44}$$

where $a_i^k(d)$ are polynomials of degree i in d. One can then apply the various existing methods for series analysis (see, e.g., Domb and Green, 1974), to obtain estimates for C_k^+ and γ_k. Indeed, we have recently used such series for these estimates, and confirmed the form $\gamma_k = \gamma + (k-2)\Delta$, which follows from Eq. (3.4) (Adler *et al.*, 1986). Some of the results are listed in Table 1. The results compare well with those from the ϵ-expansion, both for exponents and for amplitude ratios.

The low concentration method is very powerful: instead of evaluating $n_s(p)$, one may identify the polynomial $n_\Gamma(p)$, for every topological shape of a cluster Γ. To evaluate the average of any physical property A one then calculates A_Γ for the cluster Γ, and the finds the average via

$$\langle A \rangle = \sum_\Gamma A_\Gamma n_\Gamma(p) \quad . \tag{3.45}$$

If A_Γ depends only on the topological structure of the cluster, and not on its individual realizations on a d-dimensional lattice, then the dependence on d enters only via $n_\Gamma(p)$, and one easily generates series like (3.44) for $\langle A \rangle$.

In what follows we shall explicitly mention several examples. The series expansion is often the most accurate method to evaluate exponents.

4. Dilute Magnetic Systems

4.1. The dilute Ising model

Returning to Eq. (3.24), we can now study what happens at finite low temperatures. At such temperatures, the leading corrections to Eq. (3.24) will be of order e^{-2K} [coming from $S_i^\alpha = S_j^\alpha$ for $(n-1)$ replicas, and $S_i^\alpha = -S_j^\alpha$ for the remaining one]. This term breaks the symmetry of the Potts model, and results in a flow towards an alternative fixed point, at finite transition temperatures. The percolation point $p = p_c$, $T = 0$ is thus a *multicritical* point (Stauffer, 1975). Treating e^{-2K} as an additional variable near the percolation fixed point, one may ask how it rescales under the renormalization group. If $e^{-2K'} = b^\lambda e^{-2K}$, then the free energy will have the singular form

$$F_{\text{sing}} = |t|^{2-\alpha} \tilde{f}(e^{-2K}/|t|^\phi) \ , \tag{4.1}$$

with the *crossover exponent* $\phi = \lambda \nu$. The field theoretical analysis indeed confirms this form, and also yields $\phi = 1$ to all orders in ϵ (Stephen and Grest, 1977; Wallace and Young, 1978).

An alternative way to obtain this result, based on the link-node-blob picture discussed in Sec. 2.7, was proposed by Coniglio (1981): If one performs the trace over all the spins between two end spins, a linear distance b apart, then to order e^{-2K} it turns out that the only contributions to $e^{-2K'}$ will come from the singly connected bonds, the recusion relation being

$$e^{-2K'} \simeq L_{\text{sc}} e^{-2K} = b^{\tilde{S}_{\text{sc}}} e^{-2K} \ , \tag{4.2}$$

(exactly as for q, discussed at the end of Sec. 3.2). This was further discussed and elucidated by Aharony, Gefen and Kantor (1984, 1986). Thus, $\lambda = \tilde{S}_{\text{sc}} = 1/\nu$ and $\phi = \lambda \nu = 1$.

At these low temperatures, the effective coupling of the spins in the "blobs" is much stronger than along the singly connected bonds, and thus the spin correlation function decays only along these latter bonds. For two spins on the same cluster, at a distance r apart, we have

$$\langle S(0)S(r) \rangle = (\tan hK)^{L_1(r)} = e^{-L_1(r)/\xi_1} \ , \tag{4.3}$$

with the one-dimensional thermal correlation length

$$\xi_1 = (-\ln \tan hK)^{-1} \simeq \tfrac{1}{2} e^{+2K} \quad . \tag{4.4}$$

At p_c, the probability that these two spins belong to the same cluster is given by $G(r)$, Eq. (2.24). Thus, the *average* spin correlation is

$$[\langle S(0)S(r)\rangle]_{av} \sim r^{-(d-2+\eta)} \exp(-Ar^{\tilde{\zeta}sc}/\xi_1) \quad , \tag{4.5}$$

where we also replaced $L_1(r) = Ar^{\tilde{\zeta}sc}$, and neglected higher order terms. The fractional power of r in the exponential is different from the usual exponential decay, and the form (4.5) yielded novel features to the *spin structure factor*, i.e., the Fourier transform of (4.5), which deviates from the usual Lorentzian (Aharony, Gefen and Kantor, 1986).

4.2. The dilute spin glass

Returning to the Hamiltonian (3.5), we can now consider more complicated distributions of the exchange coefficients J_{ij} . An example of much current interest concerns the competition between ferromagnetic ($J_{ij} > .0$) and anti-ferromagnetic ($J_{ij} < 0$) interactions, which may yield frustrated (unsatisfied) bonds, and hence *spin-glass* ordering.

Consider the dilute case, in which each J_{ij} is distributed via

$$P(J_{ij}) = p_1 \delta(J_{ij} - J) + p_2 \delta(J_{ij} + J) + p_3 \delta(J_{ij}) \quad . \tag{4.6}$$

If $p = p_1 + p_2 < p_c$ then one has only finite connected clusters. We can now define the spin-glass susceptibility (Aharony and Binder, 1980),

$$\chi_{SG} = \frac{1}{N} \sum_{ij} [\langle S_i S_j \rangle^2]_{av} \quad . \tag{4.7}$$

At zero temperature, the thermal average $\langle S_i S_j \rangle$ amounts to averaging over the degenerate ground states. If there exist frustrated bonds then there may exist two (or more) degenerate ground states in which $S_i S_j = +1$ or -1, and thus $\langle S_i S_j \rangle = 0$. An example is shown in Fig. 8, for which the states $(S_1, S_2, S_3, S_4) = (1, 1, 1, 1), (1, -1, 1, 1), (-1, -1, 1, 1), (-1, -1, -1, 1)$ (and four equivalent states, with $S_i \rightarrow -S_i$) are all degenerate, with energy $-2J$. Averaging over these states, we find, e.g. $\langle S_1 S_2 \rangle = \tfrac{1}{2}$ instead of the ferromagnetic $\langle S_1 S_2 \rangle = 1$. Similarly, $\langle S_1 S_4 \rangle = 0$. Thus, the contribution to χ_{SG} will be 6, instead of $S^2 = 16$. Multiplying this new value by the probability per site to be on such a cluster, $n_4(p)$, and by the probability of the particular exchange coefficients which give these particular frustrated ground states,

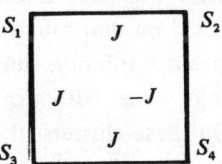

Fig. 8. An example of a frustrated plaquette in a dilute spin glass.

$4(p_1^3 p_2 + p_2^3 p_1)/p^4$, we can now sum over similar contributions from all the clusters, and obtain a series in both p_1 and p_2 for χ_{SG} (Aharony and Binder, 1980). This series diverges on a line in the p_1-p_2 plane, which signals the appearance of spin-glass ordering. On this line, $p_1 + p_2 > p_c$, since plaquettes like that shown in Fig. 8 tend to destroy the magnetic correlations between their corners. One needs less ramified clusters to recover long range order. For example, when $p_1 = p_2 = p/2$, χ_{SG} diverges at $d = 2$ only at $p_c \simeq 0.55$, and its divergence behaves as $(p - p_c)^{-\gamma_{SG}}$, with $\gamma_{SG} \simeq 2.7$.

Unlike the ferromagnetic correlations, which propagate only via the singly connected bonds, the spin-glass correlations are strongly affected by the loops, or blobs, which may destroy correlations. It remains an interesting challenge to identify exactly the geometrical property which governs the spin-glass correlations. It also remains a challenge to complete the above discussion by switching on a finite temperature, and studying the full p_1-p_2-T phase diagram (including the ferromagnetic phase near $p_1 = 1$ and the possible antiferromagnetic one near $p_2 = 1$).

4.3. The dilute antiferromagnet in a field

If all the non-zero J_{ij}'s in (3.5) are negative then the ground state of bi-partite lattices is antiferromagnetic, with the spins on two interpenetrating sublattices having opposite signs. Changing the signs of S_i on one sublattice, $\tilde{S}_i = -S_i$, thus maps the problem back onto the ferromagnetic problem, discussed above, and the order parameter (staggered) susceptibility χ is again $\sum_s n_s s^2 = \Gamma_2$.

This situation changes when a uniform field is added,

$$\mathscr{H}_h = h \sum_i S_i = \sum_i h_i \tilde{S}_i \tag{4.8}$$

with $h_i = +h$ on one sublattice and $h_i = -h$ on the other sublattice. The term (4.8) breaks the degeneracy of some of the ground states: any cluster which has more sites on one sublattice will have a *unique* ground state, with a net magnetization parallel to the field on that sublattice. Thus, $\langle \tilde{S}_i \tilde{S}_j \rangle = \langle \tilde{S}_i \rangle \langle \tilde{S}_j \rangle$, and the contribution to the susceptibility sum χ vanishes. Therefore, only clusters with an even number of sites, with exactly one half of them on one sublattice, contribute to χ. On these clusters, there are two degenerate ground states with $\langle S_i \rangle = 1$, hence $\langle \tilde{S}_i \tilde{S}_j \rangle - \langle \tilde{S}_i \rangle \langle \tilde{S}_j \rangle = 1$ and the contribution to χ is s^2. Using Eq. (3.45), we end up with

$$\chi = \sum_{\substack{s \\ \text{even}}} W\left(\frac{s}{2}, \frac{s}{2}, p\right) s^2 \quad , \tag{4.9}$$

where $W(s_A, s_B, p)$ is the probability per site to belong to a cluster of $s_A + s_B = s$ sites, with s_A on the sublattice A and s_B on sublattice B. For large s_A and s_B, one expects a random-walk-like distribution

$$W(s_A, s_B, p)/n_s(p) = 2(\pi a s)^{-1/2} e^{-(s_A - s_B)^2/(as)} \quad ; \tag{4.10}$$

hence

$$\chi \sim \sum_s n_s(p) s^{3/2} \quad . \tag{4.11}$$

Using the scaling relation (2.16) we conclude that

$$\chi \sim \xi^{\frac{3}{2}D - d} \sim (p_c - p)^{-(\gamma - \beta)/2} \tag{4.12}$$

(Aharony, Harris and Meir, 1985).

The behavior (4.12), expected at finite values of (h/J), replaces the behavior $\chi \sim \Gamma_2 \sim (p_c - p)^{-\gamma}$ for $h = 0$. As usual, one should expect a crossover behavior,

$$\chi = (p_c - p)^{-\gamma} X[(h/J)/(p_c - p)^{\phi}] \quad . \tag{4.13}$$

Since h couples to $(s_A - s_B)$, which scales with $s^{1/2}$, we can identify the rescaling of h as $b^{D/2}$, and hence $\phi = D\nu/2 = \Delta/2$. The zero temperature staggered susceptibility will thus behave as $(p_c - p)^{-\gamma}$ for $(h/J) \ll (p_c - p)^{\phi}$, and will crossover to $(p_c - p)^{-(\gamma - \beta)/2}$ for larger fields.

The terms in Eq. (4.8) may be collected into pairs, within unit cells of the antiferromagnetic state. Within such a cell, we have

$$h(S_1 t_1 + S_2 t_2) = h(S_1 t_1 - \tilde{S}_2 t_2)$$

$$= \tfrac{1}{2} h \left[(S_1 - \tilde{S}_2)(t_1 + t_2) + (S_1 + \tilde{S}_2)(t_1 - t_2) \right] \quad , \tag{4.14}$$

where $t_i = 0$ (or 1) for empty (or occupied) sites. Thus, the order parameter $(S_1 + \tilde{S}_2)/2$ feels a local *random field*, $h(t_1 - t_2)$, which has zero average. This is the basis for mapping the dilute antiferromagnet in a field onto the *random field problem* (e.g., Fishman and Aharony, 1979; Cardy, 1984). Indeed, Aharony, Harris and Meir (1985) showed that the susceptibility of the dilute random field Ising model in a field also diverges as in Eq. (4.12).

It is worth adding that in the dilute Ising model with random fields one also has the crossover exponent $\phi = D\nu/2 = (\beta + \gamma)/2$. In fact, since only spins which belong to the infinite cluster contribute to the magnetic free energy one may rewrite many of the hyperscaling relations with D replacing d with a shift in other exponents by β (Gefen *et al.*, 1982).

4.4. *Other magnetic problems*

It is clear from the above examples that different magnetic problems require different geometrical and statistical properties of the percolating clusters. There exist many more examples, and we conclude Sec. 4 with a brief partial list.

One generalization concerns Ising spins with more than two states, Potts models, $Z(N)$ models, etc. Of these, we mention only the *dilute Blume-Capel model* (Blume, 1966; Capel, 1966) with Hamiltonian

$$\mathcal{H} = - \sum_{\langle ij \rangle} J_{ij} \, S_i \, S_j + \Delta \sum_i S_i^2 \quad , \qquad (4.15)$$

where $S_i = \pm 1$ or 0, and $J_{ij} = J$ with probability p and 0 otherwise. In the pure problem, $p = 1$, one has a transition on a line $T_c(\Delta)$, which becomes first order at low T_c (and high Δ), beyond a *tricritical point*. At zero temperature, there is a first order transition at $\Delta/J = z/2$, where z is the coordination number, from $S_i \equiv 1$ to $S_i \equiv 0$. When the system is diluted, the coordination number varies among the sites, and it is no longer favorable for all the spins to flop at the same value of Δ/J. A detailed analysis (Stein and Aharony, unpublished) yields a "devil's staircase" of transitions in the Δ-p ($T = 0$) plane, which depends on details of the distribution of pieces of the infinite cluster, which are connected to the rest of it via few bonds.

Another generalization involves *long range interactions* (Stephen and Aharony, 1981; Aharony and Stauffer, 1982). If every spin has some interaction with every other spin, then at zero temperature all the spins are connected, and the percolation threshold is $p_c = 0$. However, if the interactions are not all ferromagnetic, various ordered phases may arise. In particular, if the interactions are *dipolar* then a *spin-glass* phase may replace the ferromagnetic one at low concentrations (Stephen and Aharony, 1981).

A third direction concerns *quantum effects,* which are always important at low temperatures. It is well known that a *transverse magnetic field* Γ lowers the transition temperature of the quantum Ising model. At sufficiently large Γ, $T_c(\Gamma)$ is lowered to zero. The corresponding multicritical point has in d dimensions the critical properties of a spatially anisotropic $(d + 1)$-dimensional classical Ising model (e.g., Suzuki, 1974). Dilution of the quantum model is equivalent to the introduction of randomness in d out of these $(d + 1)$ dimensions. Such a randomness, with infinite correlations along one axis, is highly relevant (e.g., Andelman and Aharony, 1985). Varying p and Γ may then reach a new $T = 0$ multicritical point, at which both the percolation and the quantum phenomena are important.

5. Conductivity and Random Walks

5.1. Scaling

As we mentioned in Sec. 2.7, one of the important properties one can measure between pairs of points on a cluster at p_c is the resistance, which scales as

$$R \sim L^{\tilde{\zeta}_R} \quad . \tag{5.1}$$

We can now repeat all the scaling arguments of Sec. 2, to obtain the following results:

(a) Equation (5.1) concerns only the *average* point-to-point resistance. We shall discuss fluctuations separately below, in Sec. 5.7.

(b) Equation (5.1) is only *asymptotic.* Generally there will appear corrections,

$$R(L) \simeq A_R L^{\tilde{\zeta}_R} (1 + a_R L^{-\omega} + \ldots) \quad . \tag{5.2}$$

(c) For a finite cluster of s bonds, one may find the resistance between any pair of two points R_{ij}, and then average, to find

$$R(s) = \sum_{i,j} R_{ij}/(2s^2) \quad . \tag{5.3}$$

One expects $R(s)$ to diverge as $s^{\tilde{\zeta}_R/D}$.

(d) On a finite sample, of size L, the average resistance between two points on the boundary is

$$\sum n_s s^2 R(s)/ \sum n_s s^2 \sim L^{\tilde{\zeta}_R} \quad . \tag{5.4}$$

(e) Below p_c, $R(L)$ decays exponentially to zero for distances $L > \xi$. For $p \neq p_c$ one expects a scaling form

$$R(L) = L^{\widetilde{\zeta}_R} r(L/\xi) \ , \tag{5.5}$$

and $r(x)$ decays exponentially for $x \gg 1$ and $p < p_c$.

(f) Above p_c, the largest cluster becomes homogeneous for $L > \xi$. It may then be described as a network of quasi-one-dimensional links (which contain blobs and dangling bonds), of linear size ξ. By the above discussion, the resistance of each such renormalized bond is $R(\xi) \sim \xi^{\widetilde{\zeta}_R}$. The resistivity of such a (hypercubic in d dimensions) network is

$$\rho \sim \xi^{d-2+\widetilde{\zeta}_R} \ , \tag{5.6}$$

and therefore its conductivity behaves as

$$\sigma \sim (p - p_c)^\mu \ , \tag{5.7}$$

with

$$\mu = (d - 2 + \widetilde{\zeta}_R)\nu \ . \tag{5.8}$$

(g) At dimensions $d > 6$, the "blobs" are negligible and thus $\widetilde{\zeta}_R = \widetilde{\zeta}_{sc} = \widetilde{\zeta}_{chem} = D_B = 2$. For $L > \xi$ one can no longer ignore the nodes, and w enters again as a dangerous irrelevant variable, which breaks the hyperscaling relation (5.8) and yields $\mu = 3$ (Aharony, Gefen and Kapitulnik, 1984; de Gennes, 1976).

5.2. *Phase transition approach*

There exist many ways to evaluate the exponent μ. These include real space renormalization group techniques, low concentration series expansions (Fisch and Harris, 1978), Monte Carlo computer simulations (e.g., Zabolitzky, 1984 and references) etc. Some of the current values are listed in Table 1.

An alternative approach uses the mapping of the resistance problem onto a q-state Potts model, in the limit $q \to 0$ (Kasteleyn and Fortuin, 1969). In the dilute case, one must consider the dilute q-state Potts model. Using replicas, as in Sec. 3.3, one must take three limits, i.e., $q \to 0$, $n \to 0$ and $T \to 0$. The exponent $\widetilde{\zeta}_R \nu$ turns out to be equal to the crossover exponent ϕ, associated with T. Although this crossover exponent is exactly equal to unity for all *finite* values of q, it has recently been shown that it has a different value for $q \to 0$ (Harris, Kim and Lubensky, 1984). For $d = 6 - \epsilon$ dimensions, Lubensky

and Wang (1986) find

$$\tilde{\mathfrak{s}}_R \nu = \phi = 1 + \frac{\epsilon}{42} + \frac{4}{3087} \epsilon^2 + \cdots \quad . \tag{5.9}$$

A third approach concerns *dilute magnets with continuous symmetry*, e.g., the dilute Heisenberg model,

$$\mathscr{H} = - \sum_{\langle ij \rangle} J_{ij} \, \mathbf{S}_i \cdot \mathbf{S}_j \quad . \tag{5.10}$$

At low temperature, the excitations of this model concern *spin waves,* and the equations for the angles between neighboring spins are practically the same as the Kirchoff equations for the voltages (Kirkpatrick, 1973). Alternatively, the low temperature renormalization group recursion relations for the exchange couplings $K_{ij} = \beta J_{ij}$ are the same as those for the conductances in the equivalent resistor network, and one finds

$$(K')^{-1} = b^{\tilde{\mathfrak{s}}_R} K^{-1} \quad , \tag{5.11}$$

so that the free energy of the dilute Heisenberg model behaves as

$$F_{\text{sing}} = |t|^{2-\alpha} \, \tilde{\mathscr{F}}(T/|t|^{\phi}) \quad , \tag{5.12}$$

with $\phi = \tilde{\mathfrak{s}}_R \nu$ (Coniglio, 1981; Aharony, 1984; Harris and Lubensky, 1984).

5.3. *Diffusion and random walks*

The *diffusion coefficient* of a random walker (e.g., on a lattice) may be defined via

$$\mathscr{D} = \frac{d}{dt} \langle r^2 \rangle \quad , \tag{5.13}$$

where $\langle r^2 \rangle$ is the mean square displacement after t time steps. Usually, $\mathscr{D}$ is a constant, and one has

$$t \sim \langle r^2 \rangle^{d_w/2} \quad , \tag{5.14}$$

with the *fractal dimensionality of the random walk* $d_w = 2$.

Following de Gennes (1976), we now turn to the notion of a *random walker on a dilute network* (the *"ant in the labyrinth"*). We start in the homogeneous regime, $L \gg \xi$. In this regime the random walk reaches a distance L practically only via the infinite cluster. The Einstein relation (Gefen *et al.*, 1983; Havlin *et al.*, 1983) now relates the diffusion coefficient on this cluster, $\mathcal{D}_\infty$, to the density of charge carriers on it, which must be proportional to P_∞, and to the d.c. conductivity,

$$\sigma \propto P_\infty \mathcal{D}_\infty \quad . \tag{5.15}$$

Using Eqs. (2.20) and (5.6), this implies that

$$\mathcal{D}_\infty \sim \xi^{-\theta} \tag{5.16}$$

with (Gefen *et al.*, 1983)

$$\theta = D - 2 + \tilde{\zeta}_R = (\mu - \beta)/\nu \quad . \tag{5.17}$$

On length scales $L < \xi$, we expect $\mathcal{D}_\infty$ to scale via $\mathcal{D}_\infty(L) = \xi^{-\theta} \tilde{\mathcal{D}}(L/\xi)$, and therefore $\mathcal{D}_\infty(L) \sim L^{-\theta}$. If the random walker is "parachuted" on the infinite cluster, then its diffusion is governed by $(d/dt)\langle r^2 \rangle \sim \mathcal{D}_\infty(r) \sim r^{-\theta}$, with $r = \langle r^2 \rangle^{1/2}$. The solution has the form (5.14), with

$$d_w = 2 + \theta = D + \tilde{\zeta}_R \quad . \tag{5.18}$$

This result applies only *on* the infinite cluster. If the random walker is parachuted at random, then we must *average over all clusters*. On a finite cluster of size R_s, we expect $\mathcal{D}(r) \sim r^{-\theta}$, i.e., $t \sim r^{2+\theta}$, for $r < R_s$, but $\langle r^2 \rangle$ can never become larger than order R_s^2. Averaging with the weights sn_s then yields

$$\langle r^2 \rangle = t\xi^{-\mu/\nu} f(t/\xi^{2+\theta}) \quad , \tag{5.19}$$

with $\langle r^2 \rangle \sim t^{(2-\beta/\nu)/(2+\theta)}$ for $t \ll \xi^{2+\theta}$ and $\langle r^2 \rangle \sim t\xi^{-\mu/\nu}$ for $p > p_c$, $t \gg \xi^{2+\theta}$. For $p < p_c$ and $t \gg \xi^{2+\theta}$ we expect $\langle r^2 \rangle \sim \langle R_s^2 \rangle \sim \xi^{2-\beta/\nu}$ (Mitescu and Roussenq, 1983).

For regular random walks, the result (5.14) arises from the distribution function $P(\mathbf{r}, t)$, which yields the probability of finding the walker at site $\mathbf{r}$ at time t. This function then assumes the Gaussian form

$$P(\mathbf{r}, t) \sim t^{-d/2} \exp(-r^2/2\mathcal{D}t) \quad . \tag{5.20}$$

It is tempting to generalize (5.20) to the percolation case on the infinite cluster, when one might expect a form (O'Shaughnessy and Procaccia, 1985)

$$P(\mathbf{r}, t) \sim t^{-d_s/2} \exp(-r^{d_w}/2\mathcal{D}_0 t) \quad , \tag{5.21}$$

with the *fracton or spectral dimension* (Alexander and Orbach, 1982; Rammal and Toulouse, 1983)

$$d_s = 2D/(2 + \theta) = 2D/d_w \quad . \tag{5.22}$$

However, the situation is somewhat more complicated. When the master equation

$$P(\mathbf{r}, t+1) - P(\mathbf{r}, t) = \sum_{\delta} [W_+(\mathbf{r}, \delta) P(\mathbf{r} + \delta, t)$$

$$- W_-(\mathbf{r}, \delta) P(\mathbf{r}, t)] \tag{5.23}$$

is diagonalized on a finite cluster, there appear eigensolutions of the form $P(\mathbf{r}, t) = e^{-\lambda t} \tilde{P}(\mathbf{r})$ with both positive and negative values of $e^{-\lambda}$. The negative ones imply *oscillations* with t, of the form $(-1)^t$. These oscillations may multiply Eq. (5.21) by factors like $[A + B(-1)^t]$, and yield oscillations in measurable quantities. In particular, we have recently found that the acceleration-acceleration correlation function of the random walker asymptotically obeys an asymptotic pure oscillation, of the form $(-1)^t$. Its magnitude decays with distance with a *superuniversal* exponent, r^{-2} (Aharony, Gefen, Meir, Nakanishi and Schofield, 1986).

5.4. Superconductor-metal mixtures

Consider now a *mixture of two conductors*, with bond conductances σ_1 and σ_2. The limit $\sigma_2 = 0$ corresponds to the dilute resistor network discussed above. When $\infty > \sigma_2 > 0$, the conductivity of the system becomes analytic at $p = p_c$, similar to the behavior of a magnet at finite magnetic fields. In analogy to Eq. (3.12) one thus writes (Straley, 1977; Efros and Shklovskii, 1976; Bergman and Imry, 1977)

$$\sigma = \sigma_1 |p - p_c|^{\mu} \mathscr{F}[(p - p_c)(\sigma_2/\sigma_1)^{-a}] \quad , \tag{5.24}$$

with $\mathscr{F}(x) \to$ const. for $x \to 0^+$. For a metal-superconductor alloy, $\sigma_2 > 0$, $\sigma_1 \to \infty$, we expect σ to diverge as $p \to p_c^-$, $\sigma \sim (p - p_c)^{-s}$, and hence $\mathscr{F}(x) \to (-x)^{-\mu-s}$ for $x \to 0^-$, and we identify $a = 1/(s + \mu)$.

The scaling form (5.24) now suggests a generalization of the "ant" into a "termite" (de Gennes, 1979), which moves quickly on the "good" conductor (σ_1) and slowly on the "bad" one ($\sigma_2 \ll \sigma_1$). There have recently been several simulations of such termites (Adler *et al.*, 1985; Bunde *et al.*, 1985; Leyvraz *et al.*, 1986). A combination of Eqs. (5.19) and (5.24) is then suggested,

$$\langle r^2 \rangle = t\sigma_2 \xi^{s/\nu} F(t\sigma_1/\xi^{2+\theta}, t\sigma_2/\xi^{2-\bar{\theta}}) \quad , \tag{5.25}$$

with

$$\bar{\theta} = (s+\beta)/\nu \quad . \tag{5.26}$$

Taking the superconducting limit, $\sigma_1 \to \infty$, the first argument in (5.25) disappears and we find

$$\langle r^2 \rangle = t\sigma_2 \xi^{s/\nu} \bar{F}(\infty, t\sigma_2/\xi^{2-\bar{\theta}}) \quad , \tag{5.27}$$

in complete analogy to (5.19).

There have been several attempts to relate the exponent s to "static" exponents, e.g., β and ν, but none of these is yet fully confirmed (for a review, see, e.g., Aharony, 1986).

5.5. AC conductivity

The arguments of Sec. 5.3 introduced a special time scale, $\tau = \xi^{2+\theta}$, at which the crossover from anomalous to normal diffusion on clusters occurs. The assumption that τ is the *only* relevant time scale led to the scaling form in Eq. (5.20). Consider now the frequency dependent ac conductivity, $\sigma(\omega)$. The above assumption implies that σ depends on ω only via the product $\omega\tau$,

$$\sigma(\omega) = \xi^{-\mu/\nu} S(\omega\tau) \quad . \tag{5.28}$$

With this assumption, the high frequency limit $\omega\tau \gg 1$ yields $\sigma(\omega) \sim \omega^x$, with $x = \mu/[\nu(2+\theta)]$. A Kramers-Kronig transformation on (5.28) then yields the dielectric constant (Gefen *et al.*, 1983),

$$\epsilon(\omega) - 1 \sim \xi^{2-\beta/\nu} E(\omega\tau) \quad , \tag{5.29}$$

with $\epsilon(\omega) \sim \omega^{-(1-x)}$ for $\omega\tau \gg 1$ and $\epsilon \sim \xi^{2-\beta/\nu}$ for $\omega\tau \ll 1$.

An alternative approach starts from Eq. (5.24), and replaces σ_2 by the imaginary "conductivity" of a dielectric insulator, $\sigma_2 = i\epsilon_2/4\pi$. Equation (5.24) then yields (Bergman and Imry, 1977)

$$\sigma = |p - p_c|^\mu \mathscr{S}(\omega\tau_2) \quad , \tag{5.30}$$

with $\tau_2 \sim (p-p_c)^{-(\mu+s)} \sim \xi^{(\mu+s)/\nu}$. This is an *additional* time scale related to the *polarizability of the medium between the clusters* (unlike τ, which describes motion *within* clusters). If τ_2 is the only relevant time scale, then $\sigma(\omega) \sim \omega^x$ with $x = \mu/(s+\mu)$ for $\omega\tau_2 \gg 1$, and $\sigma(\omega) \sim (p-p_c)^{-s}$ (Efros and Shklovskii,

1976; Bergman and Imry, 1977). In general, there is no reason *a priori* to ignore either τ or τ_2. Writing

$$\sigma(\omega) = \xi^{-\mu/\nu}\tilde{s}(\omega\tau, \omega\tau_2) \ , \tag{5.31}$$

one may expect different properties in different time regimes. In particular, the exponent x for the high frequency behavior $\sigma(\omega) \sim \omega^x$ cannot be identified without additional information. A full analysis of the frequency dependence of $\sigma(\omega)$ remains to be done.

5.6. *Fractons*

After t time steps, the random walker covers a linear distance of order $r \sim t^{1/d_w}$, and therefore a number of sites of order $S \sim r^D \sim t^{D/d_w} \sim t^{d_s/2}$, with the *fracton dimensionality* d_s defined in Eq. (5.22). Indeed, the probability to return to the origin is given by $P(0, t) \sim t^{-d_s/2} \sim 1/S$, as in Eq. (5.21).

There have been many discussions in the literature concerning the relevance of the above discussion to the motion of *phonons* (see e.g., Alexander, 1983). Assuming that the elastic energy includes terms which are isotropic in the relative displacements of atoms $[\mathbf{u}(\mathbf{r} + \boldsymbol{\delta}) - \mathbf{u}(\mathbf{r})]^2$, then the equations of motion assume a form similar to Eq. (5.23), except that on the left-hand side $[P(\mathbf{r}, t + 1) - P(\mathbf{r}, t)]$ is replaced by the acceleration, $d^2\mathbf{u}/dt^2$. The eigenvalues of this equation are thus related to the square of the phonon frequency, ω^2, and the density of states, which was of the form ω^{d-1} for homogeneous systems, now crosses over to ω^{d_s-1}, related to the Laplace transform of the probability to return to the origin $t^{-d_s/2}$. Alexander and Orbach (1982) called the related (localized) excitations *"fractons."*

The above analogy between walks and elasticity also implies that the elastic modulus of the dilute system behaves as the conductivity, i.e., decreases to zero as $p \to p_c^+$ with $B \sim (p - p_c)^\mu$. Much recent attention has been devoted to the alternative case, in which the isotropic terms are absent. In that case one still finds power laws like $B \sim (p - p_c)^T$, but the exponent T turns out to be larger than μ (Kantor and Webman, 1984; Bergman and Kantor, 1984; Feng and Sen, 1984). This also implies smaller values for d_s. We shall not dwell on these elastic models any further here.

Returning to the assumption that $\xi^{2+\theta}$ is the only relevant time scale, the density of states may be written in the crossover scaling form

$$N(\omega) = \omega^{d_s-1} N(\omega\tau) \ , \tag{5.32}$$

with $N(x) \sim x^{d-d_s}$ for $x \ll 1$. The normalization condition yields an interesting step-like increase in N around $\omega \sim 1/\tau$ (Aharony, Alexander, Entin-Wohlman and Orbach, 1985).

Much of the theoretical interest in the fracton dimensionality arose from Alexander and Orbach's observation, that $d_s \simeq \frac{4}{3}$ for percolation clusters at $d \geqslant 2$. Following Rammal and Toulouse (1983), Leyvraz and Stanley (1983) presented plausibility arguments that $d_s = \frac{4}{3}$ should hold exactly, at least at high dimensionalities. These arguments concentrate on the number of visited sites, $S(t)$, and are specific for the percolation infinite cluster at p_c. One constructs the cluster simultaneously with the motion of the "ant." At each time step, the ant may either move to one of the already visited sites, or try to move into one of G growth sites never investigated before. In the latter case, this site may either be unoccupied, whence G decreases by $\Delta G = -1$ and the number of blocked sites, B, increases by unity, or it is occupied, whence $\Delta G > 0$ and the ant moves there. If there are no correlations then G is dominated by random fluctuations: $G \propto (S + B)^{1/2} \sim S^{1/2}$. The probability to increase S per unit time is $dS/dt \propto G/S \propto S^{-1/2}$, and hence $S \propto t^{2/3}$ (see also Rammal and Toulouse, 1983).

The above argument clearly breaks down at $d = 1$: Then $S(t)$ consists of a straight segment, $G \equiv 2$, $dS/dt \propto 1/S$ and $S \propto t^{1/2}$, i.e., $d_s = d = 1$. A generalization of such a boundary effect might say that G is always on the *compact narrow surface* of S. Since $S \propto r^D$, this yields $G \propto r^{D-1} \propto S^{(D-1)/D}$ and

$$d_s = 2D/(D+1) \ , \tag{5.33}$$

yielding in particular $\theta = D - 1$ and $\mu = (d-1)\nu$ (Aharony and Stauffer, 1984; Alexander, 1983). In general, one might expect a competition between the available compact surface of $S^{(D-1)/D}$ sites, and the number of randomly added sites, of order $S^{1/2}$. This led Aharony and Stauffer to suggest that Eq. (5.33) may hold only for $D < 2$, i.e., $d \lesssim 2.1$.

Recent accurate numerical studies indicate that neither $d_s = \frac{4}{3}$ nor Eq. (5.33) holds at $d = 2$, where one observes $G \propto S^{0.49}$ (instead of $S^{1/2}$ or $S^{0.47}$) and the growth sites are not located on a narrow boundary of the visited sites (Stanley *et al.*, 1984). It has recently also become clear that, although the Alexander-Orbach "rule" $d_s = \frac{4}{3}$ holds exactly for $d > 6$, it breaks down in $d = 6 - \epsilon$ dimensions, as implied by Eq. (5.9). At the moment there exists neither a satisfactory explanation for the excellent approximation $d_s \simeq \frac{4}{3}$ nor an alternative exact relation between μ (or θ) and "static" exponents (β and ν).

Before concluding this section we emphasize again that all the above arguments apply only to percolation clusters. Neither the Alexander-Orbach nor the Aharony-Stauffer relations apply to ordered structures, like the Sierpinski gaskets (Gefen *et al.*, 1981) to the "parasite" problem (ants on lattice animals), (Wilke *et al.*, 1984; Havlin *et al.*, 1984) or to diffusion on diffusion limited aggregates (Havlin, 1984).

5.7. Quantum percolation

So far, we have considered only the conductivity of *classical* particles. Much of this discussion must be modified when *quantum* effects are taken into account. In particular, disorder is known to localize some of the particle wave functions, and thus to reduce the conductivity. To study these effects, in the context of percolation geometry, it has proved very useful to consider the *dilute tight binding Hamiltonian*

$$\mathcal{H} = \sum_i \epsilon_i |i\rangle\langle i| + \sum_{\langle ij \rangle} V_{ij} |i\rangle\langle j| \quad , \tag{5.34}$$

where $|i\rangle$ is the state on the site i of the lattice, and the nearest neighbor tunneling matrix element is $V_{ij} = V$, with probability p, or zero (Shapir *et al.*, 1982; Meir *et al.*, 1986). To eliminate degeneracies, the ϵ_i's are taken as random and small.

For small p, one may solve (5.34) on finite clusters, and study the resulting wave functions, $\psi_E(i)$. One measure of their localization on the cluster Γ is the inverse participation ratio,

$$X(\Gamma) = \sum_E \left[\sum_i |\psi_E(i)|^4 \right]^{-1} \quad , \tag{5.35}$$

which is of order s^2 if the ψ's are extended, and of order s otherwise. Averaging over clusters, one generates low concentration series for $X(p)$, and finds that $X \sim (p_q - p)^{-\gamma_q}$, with $p_q > p_c$ and $\gamma_q > \gamma$ (Shapir *et al.*, 1982; Meir *et al.*, 1986). The fact that $p_q > p_c$ indicates the destructive interference on clusters at p_c. One needs less ramified clusters to maintain an extended wave function. This is qualitatively similar to the dilute spin glass case, discussed in Sec. 4.2.

The study of dilute quantum systems allows many more interesting extensions. In particular, a magnetic flux through rings changes the interference pattern of the wave function, and may *delocalize* it. Indeed, the introduction of the field through gauge phase factors resulted in a field dependent p_q, which showed interesting oscillations (Meir *et al.*, 1986). Other properties, e.g., conductivity, remain for future studies.

5.8. Noise, fluctuations, etc.

Up to this point, we assumed that all the basic bonds in the resistor network have the same unit resistance. The asymptotic average power law (5.1) then resulted from the many possible geometries which arise when one moves two electrodes, at a fixed large linear distance L, on pairs of connected sites.

One possible way to derive Eq. (5.1) quantitatively is to construct a real space renormalization group procedures. However, unlike the recursion relation for the concentration, which was discussed in Sec. 3.2, the resistance of renormalized bonds are no longer all equal to each other. Instead of the two-delta-function distribution of r (with probability p) or (with probability $1 - p$), the repeated iteration of the renormalization group generates more and more values of r, and one ends up with a quasi-continuous distribution. For example, consider the two sites 1 and 4 in Fig. 8: they are connected with probability $p' = p^4 + 4p^3(1 - p) + 2p^2(1 - p)^2$, but the renormalized resistance now has three values, r [with probability p^4], $2r$ [with probability $4p^3(1 - p) + 2p^2(1 - p)^2$] or ∞.

Stinchcombe and Watson (1976), who first studied these distributions within approximate schemes, found that the distribution function of the conductances ($\sigma = 1/r$) approaches a scaling form, which obeys

$$P^*(\sigma) = \Lambda P^*(\Lambda\sigma) ,\tag{5.36}$$

with $\Lambda = b^{-\tilde{\zeta}_R}$. This implies that all the moments of σ scale with the basic exponent $\tilde{\zeta}_R$,

$$\langle \sigma^k \rangle \sim L^{-k\tilde{\zeta}_R} .\tag{5.37}$$

Since the distribution $P^*(\sigma)$ is not Gaussian, the same applies to central moments and to cumulants.

Equation (5.37) is similar to Eqs. (2.3) and (2.7), which also arose from the scaling of the appropriate distribution function (n_s, in that case).

In addition to the wide distributions, which arise from the variations in the connectivity between points due to dilution, it has recently been found that narrow local fluctuations in the resistances of the basic resistors may also have important implications. In particular, these fluctuations directly affect the distribution of currents through the various bonds (de Arcangelis *et al.*, 1985), and hence influence the *resistance noise* (Rammal *et al.*, 1985).

To obtain a feeling for these effects, consider the Mandelbrot-Given (1984) model, shown in Fig. 5. The resistance between the ends is given by

$$R = r_1 + r_2 + [r_3^{-1} + (r_4 + r_5 + r_6)^{-1}]^{-1} ,\tag{5.38}$$

with obvious notations. If the r_i's have a very narrow distribution, we may characterize it by the average, $\langle r \rangle$, and the mean square width, $\langle (\delta r)^2 \rangle$. Substituting $r_i = \langle r \rangle + \delta r_i$ into (5.38), expanding to second order and averaging we find (Blumenfeld and Aharony, 1985)

$$\langle R \rangle = \frac{11}{4} \langle r \rangle - \frac{3}{16} \langle (\delta r)^2 \rangle / \langle r \rangle \ ,$$

$$\langle (\delta R)^2 \rangle = \frac{149}{64} \langle (\delta r)^2 \rangle \ . \tag{5.39}$$

Using the right eigenvectors of these two recursion relations we find after l iterations

$$\langle R \rangle = \left(\langle r \rangle - \frac{33 \langle (\delta r)^2 \rangle}{335 \langle r \rangle} \right) \left(\frac{11}{4} \right)^l + \frac{33 \langle (\delta r)^2 \rangle}{335 \langle r \rangle} \left(\frac{149}{64} \right)^l \ ,$$

$$\langle (\delta R)^2 \rangle = \langle (\delta r)^2 \rangle \left(\frac{149}{64} \right)^l \ . \tag{5.40}$$

Substituting $L = 3^l$, we see that

$$\langle R \rangle = \langle r \rangle L^{\tilde{\zeta}_R} \left[1 - \frac{33 \langle (\delta r)^2 \rangle}{335 \langle r \rangle^2} \left(1 - L^{2(\tilde{\zeta}_2 - \tilde{\zeta}_R)} \right) \right] \ , \tag{5.41}$$

where $\tilde{\zeta}_R = \ln(\frac{11}{4})/\ln 3 \simeq 0.92$, while $2\tilde{\zeta}_2 = \ln(\frac{149}{64})/\ln 3 \simeq 0.77$ describes the growth in the width. The exponent $2\tilde{\zeta}_2$ thus appears as a *correction to the leading power law* in (5.1), which disappears rather slowly! The noise in this renormalized bond is governed by

$$\langle (\delta R)^2 \rangle / \langle R \rangle^2 \sim L^{-b} \ , \tag{5.42}$$

with $b = 2(\tilde{\zeta}_R - \tilde{\zeta}_2) \simeq 1.07$. Numerical estimates on the real percolation problem (Rammal *et al.*, 1985) yield $b = 1.16 \pm 0.02$.

On two-terminal structures, like the links in the node-link-blob picture, one can obtain several simple bounds on b. Writing $R' = R'(r_1, r_2, \ldots, r_N)$, one has

$$b^{2\tilde{\zeta}_2} = \sum_{i=1}^{N} \left(\frac{\partial R}{\partial r_i} \bigg|_{r_i = \langle r \rangle} \right)^2 \ . \tag{5.43}$$

On a singly connected bond, $\partial R / \partial r_i = 1$. Thus,

$$2\tilde{\zeta}_2 \geqslant \tilde{\zeta}_{sc} = 1/\nu \qquad (5.44)$$

with equality holding only if there are no blobs (Wright *et al.*, 1986; Blumenfeld *et al.*, 1986). Thus, $b \leqslant 2\tilde{\zeta}_R - 1/\nu$, and the actual value is quite close to this upper bound.

Similarly, since $1 \geqslant \partial R / \partial r_i \geqslant 0$ one can show that on such a structure one always has $b > 0$ (Blumenfeld and Aharony, 1985), and also

$$2\tilde{\zeta}_2 \leqslant \tilde{\zeta}_R \leqslant D_B \quad . \qquad (5.45)$$

Similarly relations hold for higher central moments, $\langle (\delta R)^k \rangle \sim L^{k\tilde{\zeta}_k}$.

It should be noted that the above inequalities, including $b > 0$, hold only for *two terminal* structures. On three terminal structures, like the Sierpinski gasket (Fig. 4), the distribution of the new renormalized resistances are no longer independent of each other, and correlations like $\langle \delta R_1 \delta R_2 \rangle$ are generated. This increases the parameter space in Eq. (5.39), and yields additional correction exponents. Moreover, one can no longer has $\partial R / \partial r_i > 0$. For some nonlinear resistors one even finds that $b < 0$, so that the relative width of the distribution increases (Blumenfeld and Aharony, 1985).

The series of correction exponents discussed here has also been calculated using the ϵ-expansion near six dimensions (Harris *et al.*, 1984).

In addition to narrow distributions, there has also been much recent activity on power-law distributions, in which the finite resistance r may have very high values, with probability $r^{-\alpha}$ (Ben Mizrahi and Bergman, 1981; Halperin, Feng and Sen, 1985). Here one finds exponents which depend on the exponent α.

6. Conclusions

Many applications of percolation theory were not covered here. To mention a few, this article contains no mention of directed percolation, of the Hall effect of (quantum) superconductivity, of lattice animals (large percolation clusters at low p), of quantum antiferromagnets, of viscosity in percolating gels, of diffusion of many interacting particles, etc.

Hopefully, the paper has shown that practically all the physical properties near percolation are determined by the self-similar fractal geometry, and leaves the reader with some tools to go into the wide literature or (even better) to attack some new problems.

Acknowledgments

This article is dedicated to the memory of Shang-keng Ma, who was my teacher, colleague and dear personal friend.

My understanding of percolation phenomena benefitted greatly from my collaboration with others and I am grateful to all the collaborators who appear in the list of references, and to many other colleagues, whose influence is felt and appreciated.

This article was given its preliminary form during a series of lectures presented to the critical phenomena group at the University of Oslo. I enjoyed the stimulating and warm hospitality of the whole group, and particularly the useful questions and suggestions from J. Feder, T. Jossang, E. Hinrichsen and U. Oxaal.

This paper was supported in parts by grants from the U. S.-Israel Binational Science Foundation, from the Israel Academy of Sciences and from the Israel AEC.

References

Adler, J., Aharony, A. and Harris, A. B. (1984) *Phys. Rev.* **B30**, 2832.

Adler, J., Aharony, A. and Stauffer, D. (1985) *J. Phys.* **A18**, L129.

Adler, J., Aharony, A., Meir, Y. and Harris, A. B. (1986) *J. Phys.* **A** (in press).

Aharony, A. (1980) *Phys. Rev.* **B22**, 400.

Aharony, A. (1984) in *Multicritical Phenomena,* Pynn, R. and Skjeltorp, A. T., eds. (Plenum, New York) p. 309.

Aharony, A. (1986) in *Scaling Phenomena in Disordered Systems,* Pynn, R. and Skjeltorp, A. T., eds. (Plenum, New York) p. 289.

Aharony, A., Alexander, S., Entin-Wohlman, O. and Orbach, R. (1985) *Phys. Rev.* **B31**, 2565.

Aharony, A. and Binder, K. (1980) *J. Phys.* **C13**, 4091.

Aharony, A. and Fisher, M. E. (1983) *Phys. Rev.* **B27**, 4394.

Aharony, A., Gefen, Y. and Kantor, Y. (1984) *J. Stat. Phys.* **36**, 795.

Aharony, A., Gefen, Y. and Kantor, Y. (1986) in *Scaling Phenomena in Disordered Systems,* Pynn, R. and Skjeltorp, A. T., eds. (Plenum, New York), p. 301.

Aharony, A., Gefen, Y. and Kapitulnik, A. (1984) *J. Phys.* **A17**, L197; *J. Stat. Phys.* **36**, 807.

Aharony, A., Gefen, Y., Kapitulnik, A. and Murat, M. (1985) *Phys. Rev.* **B31**, 4721.

Aharony, A., Gefen, Y., Meir, Y., Nakanishi, H. and Schofield, P. (1986) to be published.

Aharony, A., Harris, A. B. and Meir, Y. (1985) *Phys. Rev.* **B32**, 3203.

Aharony, A. and Stauffer, D. (1982) *Z. Phys.* **B47**, 175.

Aharony, A. and Stauffer, D. (1984) *Phys. Rev. Lett.* **52**, 2368.

de Alcantara Bonfim, O. F., Kirkhan, J. E. and McKane, A. J. (1980) *J. Phys.* **A13**, L247.

de Alcantara Bonfim, O. F., Kirkhan, J. E. and McKane, A. J. (1981) *J. Phys.* **A14**, 2391.

Alexander, S. (1983) in *Percolation Structures and Processes,* Deutscher, G., Zallen, R. and Adler, J., eds. (Ann. Israel Phys. Soc., Vol. 5) p. 149.

Alexander, S. and Orbach, R. (1982) *J. de Phys. Lett.* (Paris) **43**, L625.

Alexandrowicz, Z. (1980) *Phys. Lett.* **80A**, 284.

Amit, D. (1976) *J. Phys.* **A9**, 1441.

Andelman, D. and Aharony, A. (1985) *Phys. Rev.* **B31**, 4305.

de Arcangelis, L., Redner, S. and Coniglio, A. (1985) *Phys. Rev.* **B31**, 4725.

Ben Mizrahi, A. and Bergman, D. J. (1981) *J. Phys.* **C14**, 909.

Bergman, D. J. and Imry, Y. (1977) *Phys. Rev. Lett.* **39**, 1222.

Bergman, D. J. and Kantor, Y. (1984) *Phys. Rev. Lett.* **53**, 511.

Blume, M. (1966) *Phys. Rev.* **141**, 517.

Blumenfeld, R. and Aharony, A. (1985) *J. Phys.* **A18**, L443.

Blumenfeld, Meir, R. Y., Harris, A. B. and Aharony, A. (1986) to be published.

Bunde, A., Hong, D. C., Majid, I. and Stanley, H. E. (1985) *J. Phys.* **A18**, L137.

Capel, H. W. (1966) *Physica* **32**, 966.

Cardy, J. (1984) *Phys. Rev.* **B29**, 505.

Coniglio, A. (1981) *Phys. Rev. Lett.* **46**, 250.

Coniglio, A. (1982) *J. Phys.* **A15**, 3829.

Deutscher, G., Zallen, R. and Adler, J., eds. (1983) *Percolation Structures and Processes,* Ann. Israel, Phys. Soc., Vol. 5.

Domany, E., Alexander, S., Bensimon, D. and Kadanoff, L. P. (1983) *Phys. Rev.* **B28**, 3110.

Domb, C. and Green, M., eds. (1974) *Phase Transitions and Critical Phenomena* (Academic Press, New York) Vol. 3.

Efros, A. L. and Shklovskii, A. L. (1976) *Phys. Status Solidi* (b) **76**, 475.

Essam, J. W. (1980) *Rep. Prog. Phys.* **43**, 843.

Essam, J. W., Gaunt, D. S. and Guttmann, A. J. (1978) *J. Phys.* **A11**, 1983.

Essam, J. W. and Gyillym (1971) *J. Phys.* **C4**, L228.

Feng, S. and Sen, P. N. (1984) *Phys. Rev. Lett.* **52**, 216.

Fisch, R. and Harris, A. B. (1978) *Phys. Rev.* **B18**, 416.

Fishman, S. and Aharony, A. (1979) *J. Phys.* **C12**, L729.

Gaunt, D. S. and Ruskin, H. (1978) *J. Phys.* **A11**, 1369.

Gefen, Y., Aharony, A. and Alexander, S. (1983) *Phys. Rev. Lett.* **50**, 77.

Gefen, Y., Aharony, A., Mandelbrot, B. B., and Kirkpatrick, S. (1981) *Phys. Rev. Lett.* **47**, 1771.

Gefen, Y., Aharony, A., Shapir, Y. and Berker, A. N. (1982) *J. Phys.* **C15**, L801.

Gefen, Y., Aharony, A., Shapir, Y. and Mandelbrot, B. B. (1984) *J. Phys.* **A17**, 435.

de Gennes, P. G. (1976) *La Recherche* **7**, 919.

de Gennes, P. G. (1979) *J. de Phys. Lett.* (Paris) **200**, 2197.

Grossman, T. and Aharony, A. (1986) *J. Phys.* **A** (in press).

Halperin, B. I., Feng, S. and Sen, P. N. (1985) *Phys. Rev. Lett.* **54**, 2391.

Harris, A. B., Kim, S. and Lubensky, T. C. (1984) *Phys. Rev. Lett.* **53**, 743; **54**, 1088 (E).

Harris, A. B. and Lubensky, T. C. (1984) *J. Phys.* **A17**, L609.

Harris, A. B., Lubensky T. C., Holcomb, W. K. and Dasgupta, C. (1975) *Phys. Rev. Lett.* **35**, 327, 1397.

Havlin, S. (1984) *Phys. Rev. Lett.* **53**, 1705.

Havlin, S., Ben Avraham, D. and Sompolinsky, H. (1983) *Phys. Rev.* **B25**, 5828.

Havlin, S., Djordjevic, Z., Majid, I., Stanley, H. E. and Weiss, G. H. (1984) *Phys. Rev. Lett.* **53**, 178.

Havlin, S. and Nossal, R. (1984) *J. Phys.* **A17**, L427.

Hermann, J. J. and Stanley, H. E. (1984) *Phys. Lett.* **53**, 1121.

Hong, D. C. and Stanley, H. E. (1984) *J. Phys.* **A16**, L475, L525.

Houghton, A., Reeve, J. S. Wallace, D. J. (1978) *Phys. Rev.* **B17**, 2956.

Kantor, Y. and Webman, I. (1984) *Phys. Rev. Lett.* **52**, 1891.

Kapitulnik, A., Aharony, A., Deutscher, G. and Stauffer, D. (1983) *J. Phys.* **A16**, L269.

Kapitulnik, A., Frid, N. and Deutscher, G. (1984) *J. de Phys. Lett.* (Paris) **45**, L401.

Kasteleyn, P. W. and Fortuin, C. M. (1969) *J. Phys. Soc. Japan* Suppl. **26**, 11.

Kirkpatrick, S. (1973) *Rev. Mod. Phys.* **45**, 574.

Kirkpatrick, S. (1976) *Phys. Rev. Lett.* **36**, 69.

Leyvraz, F., Adler, J., Aharony, A., Bunde, A., Coniglio, A., Hong, D. C., Stanley, H. E. and Stauffer, D. (1986) *J. Phys.* **A** (in press).

Leyvraz, F. and Stanley, H. E. (1983) *Phys. Rev. Lett.* **51**, 2048.

Lubensky, T. C. and Wang, J. (1986) *Phys. Rev.* **B** (in press).

Ma, S.-k. (1976) *Modern Theory of Critical Phenomena* (Benjamin, Reading, Mass).

Mandelbrot, .B. B. (1982) *The Fractal Geometry of Nature* (Freeman, San Francisco).

Mandelbrot, B. B. and Given, J. A. (1984) *Phys. Rev. Lett.* **52**, 1853.

Margolina, A., Herrmann, H. J. and Stauffer, D. (1982) *Phys.* **93A**, 73.

Margolina, A., Family, F. and Privman, V. (1984) *Z. Phys.* **54**, 321.

Meir, Y., Aharony, A. and Harris, A. B. (1986) *Phys. Rev. Lett.* **56**, 976.

Mitescu, C. D and Roussenq, J. (1983) in *Percolation Structure and Processes*, Deutscher, G., Zallen, R. and Adler, J., eds. (Ann. Israel Phys. Soc., Vol. 5) p. 149.

Nakanishi, H. and Reynolds, P. J. (1979) *Phys. Lett.* **71A**, 252.

Nakanishi, H. and Stanley, H. E. (1978) *J. Phys.* **A11**, L189.

Nienhuis, B. (1982) *J. Phys.* **A15**, 199.

den Nijs, M. P. M. (1979) *J. Phys.* **A12**, 1857.

O'Shaughnessy, B. and Procaccia, I. (1985) *Phys. Rev.* **A32**, 3073.

Pike, R. and Stanley, H. E. (1981) *J. Phys.* **A14**, L169.

Priest, R. G. and Lubensky, T. C. (1976) *Phys. Rev.* **B13**, 4159; **14**, 5125 (E).

Pytte, E. (1979) *Phys. Rev.* **B20**, 3929.

Rammal, R., Tannous, C., Breton, P., and Tremblay, A. M. S. (1985) *Phys. Rev. Lett.* **54**, 1718.

Rammal, R. and Toulouse, G. (1983) *J. de Phys. Lett.* (Paris) **44**, L13.

Reynolds, P. J., Stanley, H. E. and Klein, W. (1980) *Phys. Rev.* **B21**, 1223.

Shapir, Y., Aharony, A. and Harris, A. B. (1982) *Phys. Rev. Lett.* **49**, 486.

Skal, A. S. and Shklovskii, B. I. (1975) *Sov. Phys.-Semicond.* **8**, 1029.

Stanley, H. E. (1977) *J. Phys.* **A10**, L211.

Stanley, H. E., Majid, I., Margolina, A. and Bunde, A. (1984) *Phys. Rev. Lett.* **53**, 1706.

Stanley, H. E., Reynolds, P. J., Redner, S. and Family, F. (1982) in *Real Space Renormalization*, Burkhardt, T. W. and van Leeuwen, J. M. J., eds. (Springer, Berlin).

Stauffer, D. (1975) *Z. Phys.* **B22**, 161.

Stauffer, D. (1979) *Phys. Reports*, **54**, 1.

Stauffer, D. (1980) *Z. Phys.* **B37**, 89.

Stauffer, D. (1985) *Introduction to Percolation Theory* (Taylor and Francis, London).

Stephen, M. J. and Aharony, A. (1981) *J. Phys.* **C14**, 1665.

Stephen, M. J. and Grest, G. S. (1977) *Phys. Rev. Lett.* **38**, 567.

Stinchcombe, R. (1986) in *Scaling Phenomena in Disordered Systems,* Pynn R. and Skjeltorp, A. T., eds. (Plenum, New York).

Stinchcombe, R. and Watson, B. P. (1976) *J. Phys.* **C9**, 3221.

Straley, J. P. (1977) *Phys. Rev.* **B15**, 5733.

Suzuki, M. (1974) *Phys. Lett.* **34A**, 94.

Skyes, M. F., Gaunt, D. S. and Ruskin, H. (1976) *J. Phys.* **A9**, 1899.

Toulouse, G. (1974) *Nuovo Cimento* **B23**, 234.

Vannimenus, J. and Nadal, J. P. (1984) *Phys. Reports,* **103**, 47.

Voss, R. F. (1984) *J. Phys.* **A17**, L373.

Wallace, D. J. and Young, A. P. (1978) *Phys. Rev.* **B17**, 2384.

Wilke, S., Gefen, Y., Ilkovic, V., Aharony, A. and Stauffer, D. (1984) *J. Phys.* **A17**, 647.

Wright, D. C. Bergman, D. J. and Kantor, Y. (1986) *Phys. Rev.* **B33**, 396.

Zabolitzky, J. G. (1984) *Phys. Rev.* **B30**, 4077.

COMPLEX ANALYTIC METHODS FOR VISCOUS FLOWS IN
TWO DIMENSIONS

D. Bensimon, L. P. Kadanoff, S. Liang, B. I. Shraiman & C. Tang

The James Franck Institute
The University of Chicago
5640 S. Ellis Avenue, Chicago, IL 60637
USA

This paper is an expository treatment of recent work on using complex analytical methods for understanding the stability of hydrodynamic flow patterns in two-dimensional or almost two-dimensional geometries. We want to know the instabilities which might arise when a more viscous fluid is displaced by a less viscous one and also how surface-tension effects can restore the stability of non-trivial flow patterns.

In the first section, we describe the physical situation, restate the description in terms of partial differential equations, and summarize our state of knowledge about the solutions to the equations and the physical phenomena that arise. In two-dimensional problems, one can often make considerable progress by using calculational methods based upon analytic functions of complex variables. Section 2 describes how these methods can be used to obtain exact solutions for zero surface tension, while Sec. 3 sets up the interface equations for nonzero surface tension. Finally, the fourth section uses complex-variable methods to describe the stabilization of finger-like flow patterns.

1. The Saffman-Taylor Problem. Where Do We Stand?

1.1. Introduction: phenomenology and the basic equations

The formation and evolution of dynamical structures is one of the most exciting areas of nonlinear phenomenology. Such pattern formation problems are common in hydrodynamic systems. Perhaps the best studied ones involve the patterns formed by the interface between two phases: a solid and a fluid or two fluids. In turn, one of the simplest problems of this class is the Saffman-Taylor (1958) problem in which two fluids move in the narrow space between two plates. This geometry is called a Hele-Shaw (Hele-Shaw, 1898) cell, see Fig. 1.1. When the plate separation, b, is very small the problem is effectively two-dimensional. If we call the coordinates perpendicular to the plates z, and the other two x and y, we can specify the problem by the two components of the velocity, v_x and v_y, the pressure, $P(x, y)$ and a two-component vector $\gamma(s)$ which sweeps out the position of the interface as s is varied.

The basic equations involved are very simple indeed. In each fluid, the average velocity parallel to the plates is proportional to a local force (Saffman and Taylor, 1958)

$$\mathbf{v}(x, y) = -K_i [\nabla P(x, y) - \rho_i \mathbf{g}] \ . \tag{1.1}$$

Here $i = 1, 2$ labels the different fluids, ρ_i is the density, and the constant K_i is given in terms of the fluid viscosity, μ_i, and the plate-spacing, b, as

$$K_i = \frac{b^2}{12\mu_i} \ , \tag{1.2}$$

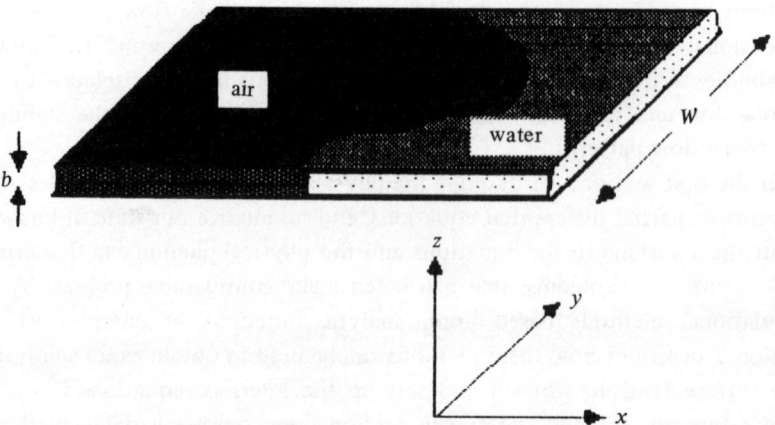

Fig. 1.1. A sketch of a Hele-Shaw cell and our coordinate system.

while g is the component of the gravitational acceleration parallel to the plates. Equations (1.1) and (1.2) constitute the Darcy approximation. They are derived in a trivial way from the Navier-Stokes equation by considering a parabolic flow profile parallel to the plates with a velocity which vanishes at both plates. Then, v is the average over the perpendicular direction of the actual velocity.

The remaining equations are easy to write down. Assume that the fluids are incompressible so that the divergence of the velocity vanishes. Then, in each fluid

$$\nabla^2 P = 0 \ . \tag{1.3}$$

Continuity also implies a boundary condition that, at the interface, the normal components of the velocity are equal to each other and to the speed of the interface:

$$v_n = - K_1 (\nabla P_1)_n = - K_2 (\nabla P_2)_n \ , \tag{1.4}$$

where the gradients are evaluated at the points $\gamma(s)$.

One more boundary condition is needed to give the jump in pressure across the interface. Theorists working on this problem often choose to take the pressure jump to be the surface tension, T, times the curvature, κ, observed in the x-y plane, i.e.,

$$\Delta P = T\kappa \ . \tag{1.5a}$$

This formula would follow were the classical Gibbs-Thomson equations for the pressure jump really applicable. This, in turn, would be true were the Hele-Shaw cell really two-dimensional. But, the cell lives in a three-dimensional world in which there are two radii of curvature for the surface. The larger one, R, has the smaller effect upon the pressure drop, while the smaller one (which is roughly $b/2$) dominates. Hence Eq. (1.5a) makes very little experimental sense. Park and Homsy (1984) suggested an alternative boundary condition which might better describe a situation in which a fluid which wets the plates is displaced by one which does not. From an asymptotic analysis they derived an expression for the pressure jump that one should use instead of Eq. (1.5a),

$$\Delta P = \frac{T}{b/2} \left[1 + 3.80 \left(\frac{\mu v_n}{T} \right)^{2/3} \right] + \frac{\pi}{4} T\kappa \ . \tag{1.5b}$$

The first term in Eq. (1.5b), $2T/b$, is independent of x and y so it does not really affect the motion. The other two terms act together, producing forces which tend to flatten out the interface.

We shall look at the simplest possible situation. Let the plates be very long rectangles with width W (Fig. 1.1). The second fluid is, say, air so that we may take it to have a negligible density and viscosity. Let the first fluid be, say, water and let the air be pushing it so that it moves with an average velocity, U. The plates are horizontal so that gravity does not enter.

The sidewalls are rigid. This is represented by using a slip boundary condition so that at the sidewalls the normal component of the velocity vanishes

$$(\mathbf{v})_n = 0 .$$
(1.6)

This free slip boundary condition may not be realistic for the true experimental situation, but it may be an essentially correct approximation for small b, where there is a boundary layer of width b near the surface over which $\mathbf{v}$ may vary quite rapidly.

Now, both in experiments and simulations, one observes three different types of motion:

Case A: Small U or negative U. (The latter implies that the water is pushing on the air.) An initial interface which perhaps has a few bumps in it eventually flattens out and forms a straight boundary between the two fluids.

Case B: Intermediate U. Any initially present bump grows and forms a stable finger (Fig. 1.2). The width of the finger is a multiple, λ, of the channel width W and varies with velocity. Under the stated conditions, in which we can neglect the second fluid, there is only one dimensionless parameter[1] entering these equations, namely

$$d_0 = \frac{\pi^2}{3} \frac{b^2}{W^2} \frac{T}{\mu U} ,$$
(1.7)

which, therefore, acts as a control parameter. Here U is the fluid velocity in the region far downstream from the finger. We call d_0 the "surface tension parameter." The dependence of the finger width on d_0 is an interesting quantity to predict theoretically. Below, we discuss our results for this dependence and those of others.

Case C: Large U. This corresponds to the parameter d_0 being very small. In this domain, several types of time-dependent behavior may be observed. For the

[1] Since there is no agreement on the exact form of the surface-tension parameter we will, for the convenience of the reader, relate our parameter d_0 to the one used by other workers in the field. Thus the parameter κ used by McLean and Saffman (1981) is $\kappa = d_0 \lambda/(1-\lambda)^2$. The parameters B, τ introduced respectively by Tryggvason and Aref (1983) and DeGregoria and Schwartz (1985) are $\tau = B = d_0/(2\pi)^2$.

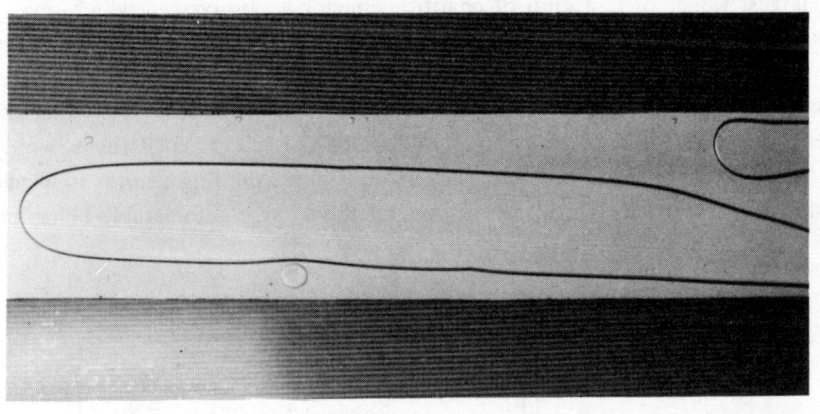

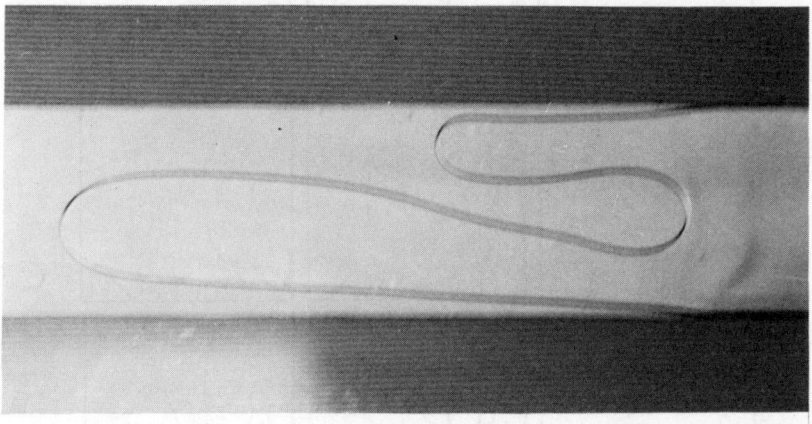

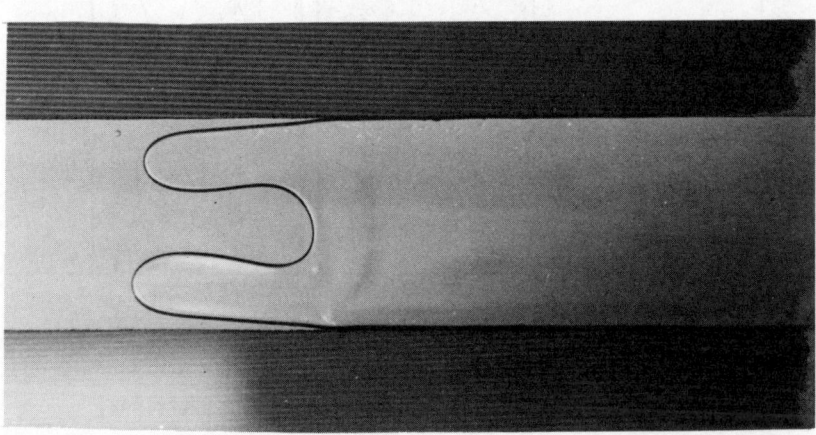

Fig. 1.2. Competition between two bumps leading to the emergence of a single propagating finger; courtesy of Tabeling and Libchaber (1985).

very largest values of U a kind of chaotic behavior is observed in which several fingers are formed which may branch and split (Fig. 1.3). There is a tendency for the tallest fingers to get ahead and leave the smaller ones well behind. Thus there is essentially a cascade into large length scales, which saturates when the fingers become of width comparable to that of the cell. Even if there is only one finger in the channel, for these large values of d_0 the finger tends to wiggle up and down, partially split, and in general show quite an unstable behavior.

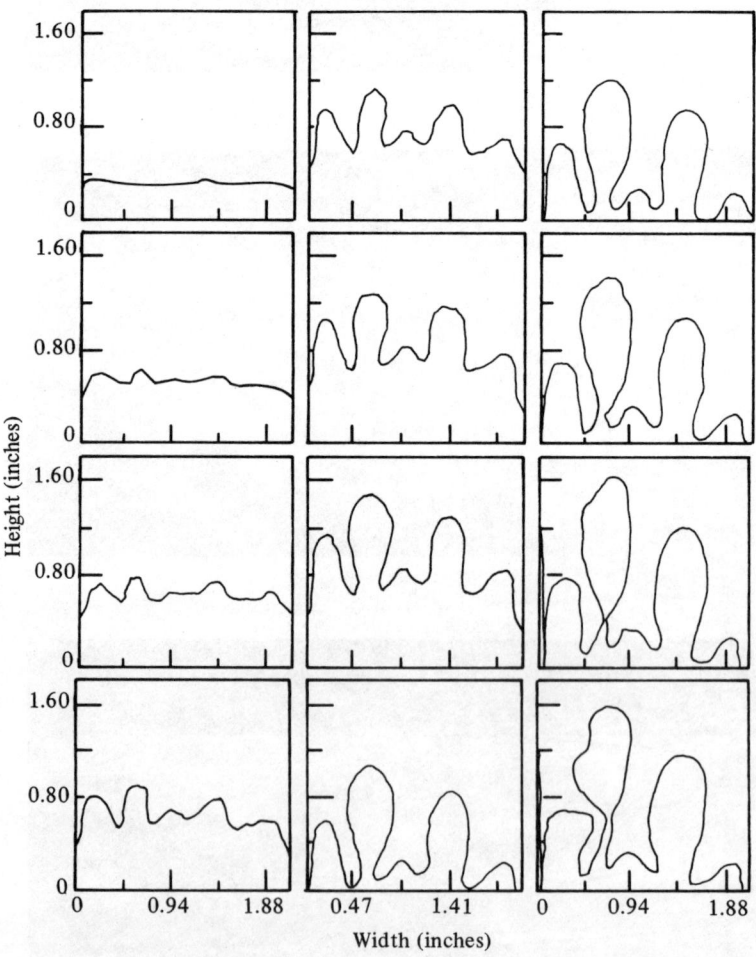

Fig. 1.3. Chaotic behavior in a Hele-Shaw cell; courtesy of Maher (1984).

1.2. Stability analysis (Chuoke, et al., 1959)

The first step is to look at the stability of an almost-flat interface. Let the flat interface be at a position $x(y) = Ut$, which moves with velocity U relative to the walls. A small deviation from flatness may be represented by writing

$$x(y) = Ut + A(t)\cos qy \; , \tag{1.8a}$$

where $A(t)$ is considered to be small. If A vanished, the velocity U would be produced by a pressure gradient $-U/(b^2/12\mu)$. If we add to this zero-order term a term produced by the deviation from flatness we find a result like (1.8a), namely

$$P(x,y) = P_0 - \frac{U}{(b^2/12\mu)}(x - Ut) + B(x,t)\cos qy \; .$$

For $P(x,y)$ to obey Laplace's equation $B(x,t)$ must vary as e^{qx} or e^{-qx}. The former is impossible if the pressure is to remain finite as $x \to \infty$. Thus, we find

$$P(x,y) = P_0 - \frac{U}{b^2/12\mu}(x - Ut) + B(t)e^{-qx}\cos qy \; , \tag{1.8b}$$

where P_0 is constant. The boundary condition at the sidewalls then requires the wavevector q to be

$$q = \frac{2\pi n}{W} \; , \tag{1.9}$$

where n is a positive integer.

If A is small, Eq. (1.8a) gives the velocity of the interface to be

$$U_n = U + \dot{A}(t)\cos qy \; .$$

On the other hand, the Darcy equation (1.1) and pressure equation (1.8b) together imply

$$U_n = U + \frac{b^2}{12\mu}qB(t)\cos qy \; ,$$

if we neglect terms of order A^2 or AB. In this way, we derive two equations for U_n and thus get one relationship between A and B, namely

$$\dot{A}(t) = B(t)\frac{b^2}{12\mu}q \; . \tag{1.10}$$

The final relationship is derived by calculating the terms in the pressure jump which are proportional to $\cos qy$, using the fact that pressure in the air is constant. On one hand from Eqs. (1.8a) and (1.8b), this part of the pressure jump is

$$\Delta P(y) = \left(\frac{U}{b^2/12\mu} A - B \right) \cos qy \ . \tag{1.11}$$

On the other hand Eq. (1.5a) gives the pressure jump as

$$\Delta P(y) = -T\kappa \approx -T \frac{d^2}{dy^2} x(y) \ . \tag{1.12}$$

Putting this result together with equations (1.10) and (1.11) one finds that A satisfies

$$\dot{A} = A \left(\frac{U}{b^2/12\mu} - Tq^2 \right) \frac{b^2}{12\mu} q \ . \tag{1.13}$$

This result is easily interpreted. First of all, notice that when the quantity in parentheses is positive the flat interface is unstable, when it is negative the interface is stable against a disturbance of the given wavenumber. Since, according to Eq. (1.9) the minimum value of q is $2\pi/W$, the flat interface will be unstable against some perturbation whenever the d_0 defined by Eq. (1.7) obeys $d_0 < 1$. Alternatively expressed, a very long interface will be unstable against a perturbation of wavelength $l = 2\pi/q$ whenever $l > W\sqrt{d_0}$. Hence for very small surface tension, the system will be unstable against even very short-wavelength perturbations.

The physical source of this instability lies in the geometry of the moving interface. Imagine a situation in which the pressure difference along the length of the channel is fixed. Then, since the pressure in the air is constant, the larger gradients in the pressure appear at the end of the largest fingers of air. Hence these fingers move faster than the rest. Hence they get further ahead. The entire system, is in this way, destabilized by the motion.

Conversely, the surface tension tends to stabilize and smooth out the smallest fingers, those with a radius of curvature less than $W\sqrt{d_0}$. These smallest fingers have at their ends a large pressure drop from across the air-water interface. Water flows in toward these low pressure regions, pushing the smallest fingers backward. Hence they are smoothed out by the surface tension.

Something very peculiar happens when the surface tension goes to zero. Then the most unstable wavelength becomes shorter and shorter. In the limit, $T \to 0$, the shortest wavelengths are the most unstable. One suspects that, in this limiting case the entire physical problem may well be poorly defined.

1.3. The small surface tension puzzle

Let's for a moment ignore the problem of the short wavelength instability in the absence of surface tension and ask about the steady states only. In their classic paper, Saffman and Taylor (1958) found a one parameter family of finger shaped steady state solutions. These solutions correspond to different values of λ, the ratio of the finger width to the width of the cell. The finger shape is described by the formula to be given in Sec. 2,

$$\frac{x}{W} = \frac{(1-\lambda)}{\pi} \ln \cos\left(\frac{\pi y}{\lambda W}\right) . \tag{1.14}$$

These shapes seemed to be quite similar to those that were observed experimentally. However, there were two serious problems. First, in the experiment, a finger of a well defined width was observed at each given velocity. The zero surface tension theory could not predict that, since by varying λ one could obtain fingers of any width at all. Second, from our analysis above, one should expect the $T = 0$ solutions to be completely unstable, while the fingers that were observed were quite stable. Both problems were related to the singular nature of the zero surface tension limit. Another point made by Saffman and Taylor in their paper was that no fingers with λ less than $\frac{1}{2}$ were seen in the experiment at all, and they asserted that $\frac{1}{2}$ was the asymptotic width of the finger in the $d_0 \to 0$ limit.

The question of "velocity," or finger width λ, selection was again taken up by McLean and Saffman (1981) who looked for the steady state solutions in the presence of a small but finite surface tension. Numerically solving the integral equation for the interface, they found a unique solution for a given value of the surface-tension parameter, rather than a one parameter family. Their work was further extended by Vanden-Broeck (1983) who found not just one, but a discrete set of solutions. However, for all of his solutions, as d_0 goes to zero, λ goes to one half.

Thus, the "degeneracy" of the steady states is lifted by the effects of the surface tension, which is a *singular* perturbation in this problem (see Bender and Orsag (1978) for a discussion of singular perturbations). This phenomenon, common to a large class of nonlinear problems arising in physics, was studied by Barenblatt and Zel'dovich (1972) (see also Barenblatt, 1977) in the general context of similarity solutions to partial differential equations. (The propagating solution, such as the Saffman-Taylor finger, may be thought of as a kind of similarity solution as well.) Barenblatt and Zel'dovich point out that in cases where singular perturbations are involved, the search for the similarity solutions leads to nonlinear eigenvalue problems. These eigenvalues then determine the scaling, or in case of propagation, the velocity, of the similarity solution. The existence of a continuous family of solutions would then correspond to a continuous spectrum. More commonly, a discrete spectrum is found. Thus, the work of McLean and Saffman and Vanden-Broeck fit nicely into this general[2] framework.

The results of McLean and Saffman (1981) and Vanden-Broeck (1983) produced a theoretical prediction for the finger-width dependence on the control parameter. Alas, the stability problem remained unresolved since the analysis performed by McLean and Saffman found that the fingers remained unstable even in the presence of surface tension! A result contradictory to the experiment and numerical simulations. Hence, the point about stability remained open.

All of these difficulties arise from the subtlety of the zero surface tension limit, which is singular indeed. In fact the short wavelength instability leads to the appearance of finite time singularities in the dynamical equations for a large class of initial conditions as was shown by Shraiman and Bensimon (1984) and Sarkar (1984) (see also the work of Meyer (1982) and Howison (1985a)). These singularities correspond to $\frac{2}{3}$ power cusps in the interface. After the appearance of the cusp the calculations (and probably the solutions) break down. Some of the time dependent solutions evolving into such cusps can be found explicitly (Meyer, 1982; Shraiman and Bensimon, 1984 and Howison, 1985a). While many initial conditions lead to cusps, there are also some special initial conditions which give instead $\lambda = \frac{1}{2}$ steady fingers. These $T = 0$ results can be derived using the conformal mapping method which we shall describe in Sec. 3.

In the last year, as a result of the experimental studies of Tabeling and Libchaber (1985) and theoretical work of Kessler and Levine (1985), DeGregoria and Schwartz (1985) and Bensimon (1985) a new understanding of the stability problem began to emerge (see also the earlier experiments of Aribert (1970), as well as the more recent work of Maher (1985)). First of all, experimentally the fingers at high velocity (small d_0) *are* unstable (a fact that was observed, but for some reason ignored in Saffman and Taylor (1958)). Naively, one can try to explain this by noting that for small d_0 the unstable wavelength $l = W\sqrt{d_0}$ is much shorter than the characteristic curvature and width of the finger. Thus

---

[2] There is reason to believe that some of the other puzzling "selection" problems will be resolved along the same lines: for example, recently Pomeau and Pelce (1985) derived a nonlinear eigenvalue equation governing the shape and velocity of a dendrite (in the low Peclet number limit of the "two-sided" model). The work of Barenblatt and Zel'dovich also largely anticipated the "microscopic solvability" principle put forward by Kessler, Koplik and Levine (1984) and Ben-Jacob and coworkers (1984) to explain the growth velocities in their models of solidification.

on the length scale l the finger appears to be essentially flat and therefore should be unstable. If this were indeed the case one would expect the finger to become unstable at $d_0 \approx 1$, that is, shortly after the "primary" instability which lead to the appearance of the finger in the first place. Instead, the instability is observed at $d_0 = 10^{-2}$ and a different scenario is required. Kessler and Levine (1985) suggested that the interaction of the finger with the rigid walls makes the finger stable with respect to infinitesimal perturbations (contrary to the result of McLean and Saffman (1981)). This was corroborated by the observation by DeGregoria and Schwartz (1985) that in the numerical simulations the disturbances generated at the tip decay in amplitude as they are subvected along the side of the finger. DeGregoria and Schwartz (1985) and Bensimon (1985) then proposed that the experimentally observed behavior is due to a finite amplitude instability. Furthermore, from numerical stability analysis and simulations, Bensimon found that the noise amplitude required for destabilization decreases rapidly with d_0, and is consistent with the expression

$$\ln (noise) \sim - d_0^{-1/2} \ , \qquad\qquad\qquad (1.15)$$

(see Sec. 4.3 below). He also found the most unstable modes, which are in excellent agreement with the experiments of Tabeling and Libchaber (1985).

The physical mechanism that is involved here appears to be very similar to the one proposed earlier by Zel'dovich and coworkers (1980) in connection with the stability of cellular flames. The growth rate of a disturbance is proportional to the normal velocity of the interface, so that it is large at the tip and approaches zero toward the side of the fingers. As the finger moves forward, the disturbance moves more slowly than the tip, so that it gradually moves toward less unstable regions. When the instability becomes weak, even a little surface tension is sufficient to damp out the disturbance. The dependence of the finger width λ on the control parameter d_0 is shown in Fig. 1.4. While the dependence is similar to that observed in the experiment of Saffman and Taylor (1958), the direct comparison is not quite satisfactory. The reason is that the theorists have simplified the problem by assuming that the pressure jump on the interface is velocity independent, Eq. (1.5a), rather than a more appropriate condition given by Eq. (1.5b). Patrick Tabeling has argued that the latter, more complicated boundary condition can be heuristically incorporated as an effective surface tension parameter. In the physically important region it differs by about a factor of two from the actual surface tension of the oil used in the experiment. With this correction the theoretical predicted finger width agrees with his experimental data to within 25%, which is reasonably good in view of all the complications of the actual flow: the three-dimensional effects at the meniscus, the wetting film that the finger leaves behind, etc. Figure 1.5 shows that the theoretical obtained finger shape does agree reasonably well with the experimental data.

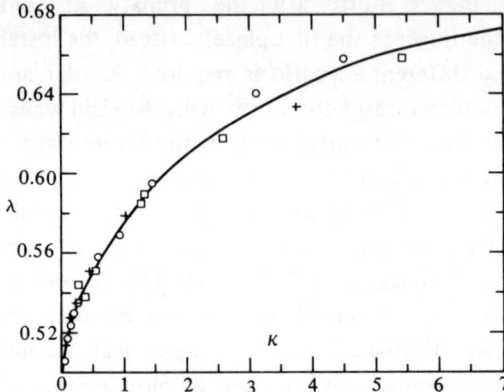

Fig. 1.4. The finger width λ as a function of the control parameter κ ($\kappa = \lambda d_0 (1-\lambda)^{-2}$). Solid line is the McLean-Saffman (1981) numerical result. Circles are Bensimon's (1985) simulations. Crosses are from the simulations of DeGregoria and Schwartz (1985). Boxes are from Liang's (1985) random walk simulations.

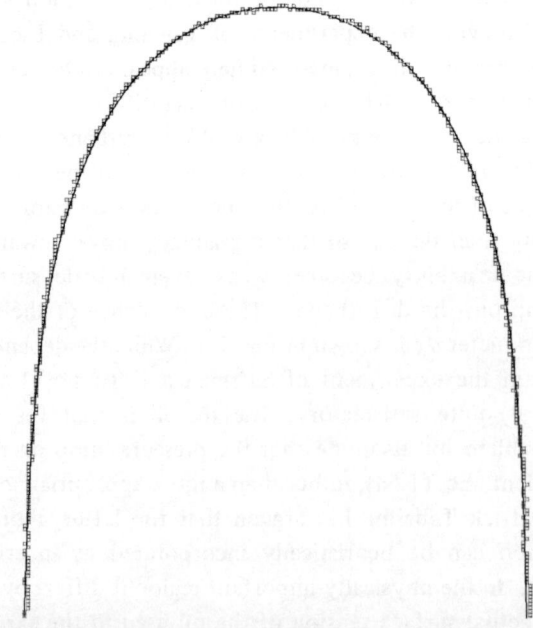

Fig. 1.5. Comparison of the finger shape obtained from simulations and experiment. The boxes are from a random walk simulation, Liang (1985), the solid line is Pitts' (1980) phenomenological scaling hypothesis which agrees with the experimental data of Saffman and Taylor (1958).

2. The Hodograph Method

2.1. *Complex analytical methods*

In this section we summarize what is known about the Saffman-Taylor problem at zero surface tension. We shall describe a method, involving analytic functions of complex variables which enables one to study the dynamics of the interface evolution in the absence of surface tension.

In the first section the Saffman-Taylor problem was formulated mathematically in terms of a Laplace equation for the pressure (or velocity potential) with two boundary conditions. As is usual in two-dimensional problems involving the Laplace equation, one can be helped considerably by using complex variable techniques (Birkhoff, 1952; Carrier, Krook, Pearson, 1966). The basic idea is to think of the velocity potential, $\phi(x, y)$ as the real part of a complex field

$$\Phi(x, y) = \phi(x, y) + i\psi(x, y) . \tag{2.1}$$

The demand that $\nabla^2 \phi = 0$ can be automatically satisfied by simply requiring that Φ be an analytic function of the complex variable

$$z = x + iy . \tag{2.2}$$

For reasons which will become more obvious later, it is better to invert the functional dependence and describe not how Φ is determined by z but rather that the potential Φ defines where we are in space. In symbols

$$z = f_t(\Phi) . \tag{2.3}$$

The subscript t indicates that the dependence changes in time. In fact we shall use the boundary conditions (1.4) and (1.5) to define the time-dependence of f_t.

Equation (2.3) is the central equation in the hodographic method, widely used in fluid mechanics. For the Saffman-Taylor problem the hodograph technique has been used to obtain a partial solution in the absence of surface tension (see Secs. 2.3 and 2.4 below), to do simulations (as discussed in Sec. 3 below), and to discuss the stability of small surface tension solutions (see Sec. 4 below).

2.2. *Basic equations*

The real velocity potential $\phi(x, y)$ is defined so that its gradient is the velocity vector. The corresponding statement for the complex case is that

$$v_x(x, y) - iv_y(x, y) = \frac{d}{dz} \Phi(z) . \tag{2.4}$$

Equation (2.4) can then be re-expressed to give several useful boundary conditions. Since v_y vanishes on the side walls, $y = \pm W/2$, the imaginary part of Φ must be constant on each wall. Denote the velocity downstream by U. Since v_y vanishes far downstream, as $\mathrm{Re}(z) \to \infty$, $\Phi \to Uz + \mathrm{const}$. Take the constant to be real and then notice that $\mathrm{Im}\,\Phi = \pm UW/2$ on the two walls. What we have just said can be converted into boundary conditions upon the unknown function f_t of Eq. (2.3), i.e.,

$$\text{if}\quad \mathrm{Re}\,\Phi \to \infty\,,\quad \text{then}\quad f_t(\Phi) \to \frac{\Phi}{U} + C_0 \quad \text{with}\quad \mathrm{Im}\,C_0 = 0\,, \qquad (2.5a)$$

$$\text{if}\quad \mathrm{Im}\,\Phi = \pm UW/2\,,\quad \text{then}\quad \mathrm{Im}\,f_t(\Phi) = \pm W/2\,. \qquad (2.5b)$$

Finally, at zero surface tension, the pressure and hence the velocity potential is constant on the interface between the two fluids. Thence, if we take this constant to be zero, we can choose Φ to be of the form *is* for s real, on the interface.

To define the interface precisely we must determine a curve for each value of the time t. One way of doing this is to define a complex function of two real variables, s and t, i.e., $\gamma(s, t)$. For each value of t, as s sweeps over its entire range, $\gamma(s, t)$ sweeps over a set of points $z_t(s)$. These points are the complex variables which give the values of $x + iy$ for all x and y on the interface. This function, $\gamma(s, t)$, is thus the solution to our problem. The third boundary condition is that, in the case of zero surface tension,

$$\gamma(s, t) = f_t(is)\,,\qquad s \in \left(\frac{-UW}{2}, \frac{UW}{2}\right)\,. \qquad (2.5c)$$

Notice that the three boundary conditions (Eqs. 2.5) have defined the physical region of Φ to be the strip:

$$\Phi = \phi + i\psi\,,\qquad -\frac{UW}{2} \leqslant \psi \leqslant \frac{UW}{2}\,,\qquad 0 \leqslant \phi \leqslant \infty\,. \qquad (2.6)$$

Within this strip, f_t must be analytic and its derivative must be nonzero. These two conditions together ensure that $\nabla^2 \phi = 0$ in the region filled by water.

One more condition must be fulfilled: the interface must move with the same velocity as the fluid. Consider the time-dependence of $\gamma(s, t)$. For each value of s, $\gamma(s, t)$ specifies the value of $x + iy$ for some point on the interface at time t. An infinitesimal time interval later, this same piece of fluid will have moved forward by an amount $(v_x + iv_y)dt$. In this way, we find one term in the time derivative of γ:

$$\frac{d}{dt}\gamma(s,t) = v_x + iv_y \; . \tag{2.7a}$$

Using Eq. (2.4), we can rewrite (2.7a) as

$$\frac{d}{dt}\gamma(s,t) = \left(\frac{d\Phi}{dz}\right)^* = \frac{1}{(dz/d\Phi)^*} = \frac{-i}{\partial_s \gamma(s,t)^*} \; . \tag{2.7b}$$

The total derivative on the left-hand side of (2.7b) represents the possibility that, as the front advances, the value of the parameter s labeling a particular piece of fluid might change. If it does at a rate ds/dt then we must add to Eq. (2.7b) a term reflecting this change to obtain

$$\partial_t \gamma(s,t) = -i\,\frac{\partial_s \gamma(s,t)}{|\partial_s \gamma(s,t)|^2} - \frac{ds}{dt}(s,t)\,\partial_s \gamma(s,t) \; . \tag{2.8}$$

Equation (2.8) will determine the motion of the interface. At first sight, it does not look as if the derivation of Eq. (2.8) is a huge amount of progress. For one unknown function, $\gamma(s,t)$ we have traded another, ds/dt. However, in fact, a considerable advance has been made. We know that $\gamma(s,t)$ is analytic and has a nonzero derivative for all values of s in the strip (2.5c). This analyticity essentially determines the parameterization of $\gamma(s,t)$. In turn the analyticity, plus the condition that ds/dt is real fully determines the solution to Eq. (2.8).

There are several different methods for obtaining solutions to Eq. (2.8). The easiest is to eliminate ds/dt by multiplying by the complex conjugate of $\partial_s \gamma$ and then taking the imaginary part of the result, to find

$$\partial_t \gamma(s,t)\,\partial_s \gamma(s,t)^* - \partial_t \gamma(s,t)^*\,\partial_s \gamma(s,t) = -2i \; . \tag{2.9}$$

Equation (2.9) plus the statement that $\partial_s \gamma(s,t)$ is analytic and nonzero whenever s lies in the strip (2.5c), yields the evolution of the interface.

2.3. *An example*

To show what all this means, we develop an example analogous to the stability analysis of Sec. 1.2, but now specific to the zero surface tension case. We shall now get not just an expansion for small amplitudes but an exact solution for the interface. To derive this, replace the guesses, (1.8), about the form of the pressure and the interface by a corresponding guess for $f_t(\Phi)$,

$$f_t(\Phi) = C_0(t) + \Phi/U + C_1(t)\,e^{-q\Phi/U} \; . \tag{2.10}$$

Here the first two terms give a flat interface while the third represents a "correction" with wave vector q. The actual interface is given by the curve traced out by γ as a function of s, where

$$\gamma(s, t) = C_0(t) + is/U + C_1(t)\, e^{-isq/U} \,. \tag{2.11}$$

Notice that the nontrivial dependence upon Φ and s is given in terms of

$$\omega = e^{-q\Phi/U} = e^{-isq/U} \,. \tag{2.12}$$

Our boundary conditions insist that C_0 is real. Choose C_1 to be real also.

To see that (2.11) is an exact solution, simply differentiate and substitute into Eq. (2.9). The result is

$$(\dot{C}_0 + \omega \dot{C}_1)(1 - C_1 q \omega^{-1}) + (\dot{C}_0 + \omega^{-1} \dot{C}_1)(1 - C_1 q \omega) = 2U \,. \tag{2.13}$$

The expression (2.11) will be an exact solution if we can insure that (2.13) is satisfied for all s. This will, in turn, be true if we can make the coefficients of ω^j for $j = 0, \pm 1$ each vanish. The resulting differential equations for C_0 and C_1 are

$$\frac{d}{dt}(2C_0 - C_1^2 q) = 2U \,, \tag{2.14a}$$

$$\frac{d}{dt}\ln C_1 = q\frac{dC_0}{dt} \,. \tag{2.14b}$$

If $C_1 = 0$ at time zero, it remains zero. Then C_0 increase linearly in time, $C_0 = U(t - t_0)$ and correspondingly the interface moves forward with speed U. If, however, C_1 starts out positive but small it will continually grow larger. The solution will remain acceptable until

$$\frac{\partial}{\partial s}\gamma(s, t) = \frac{i}{U}[1 - qC_1(t)\omega] \,, \tag{2.14c}$$

vanishes at some s. This will happen at some finite time when C_1 becomes equal to $\pm q^{-1}$. At this time the interface acquires a cusp and after that the evolution is not defined.

2.4. Finger solution

There exist a few special solutions which do not go to cusps. One kind is the family of finger solutions found by Saffman and Taylor (1958). Here if the finger has a width λW and the fluid moves with speed U at $x = +\infty$, then the speed at which the interface advances is U/λ. Thence as $\operatorname{Re}\Phi \to \infty$, we can expect a solution of the form

$$f_t(\Phi) = Ut/\lambda + \Phi/U + g(\Phi) \;, \tag{2.15a}$$

where g vanishes rapidly as $\operatorname{Re}\Phi \to +\infty$. In order to construct a finger, we need a singularity as Φ goes to the corners of the strip, which lie at $\Phi = \pm iUW/2$. One guess is that there is a logarithmic singularity at these points, i.e., that

$$g(\Phi) = \alpha \ln(1 + e^{-2\pi\Phi/UW}) \;. \tag{2.15b}$$

Given this guess, one finds that

$$\frac{\partial}{\partial s}\gamma(s,t) = \frac{i}{U} - \frac{2\pi i \alpha}{WU}\frac{1}{1 + e^{2\pi is/(WU)}} \;. \tag{2.16}$$

Then, a brief calculation based upon Eq. (2.9) shows that (2.15a) is indeed a solution if

$$\alpha = \frac{W}{\pi}(1-\lambda) \;. \tag{2.17}$$

This solution gives the profile described in Eq. (1.14), i.e., a single finger with width λ.

There are indeed other solutions of the form

$$f_t(\Phi) = Ut/\lambda + \Phi/U + \sum_{j=1}^{m} a_j \ln(e^{iq_j} + e^{-2\pi\Phi/UW}) \;, \tag{2.18}$$

with a_j real and positive and q_j real. These solutions have m "channels" going off to $x \to -\infty$, and are thus a generalization of the original Saffman-Taylor solution (2.15). These solutions are singular in the sense that the interface extends to infinity, the mapping has logarithmic singularities on the unit circle. The form of the steady-state solution given in Eq. (2.15) suggest the following ansatz for the time dependent solutions:

$$f_t(\Phi) = \Phi/U + C_0(t) + \sum_{j=1}^{m} a_j \log(\omega - p_j(t)) \;, \tag{2.19}$$

where ω is determined as a function of the complex potential Φ

$$\omega = e^{-2\pi\Phi/UW} \;. \tag{2.20}$$

The singularities of this map, $p_j(t)$ move with time but are confined outside the physical domain: $|p_j(t)| \geqslant 1$. When one of them hits the circle a channel going to $x = -\infty$ is formed. We can write down the evolution equation for these singularities. It turns out, however, to be more convenient to track the zeroes of $\partial_\Phi f$ instead. They must also lie outside the disk — otherwise the conformality requirement is not satisfied. We have

$$\partial_\Phi f = \alpha_0 \frac{\prod_{j=1}^{m} [\omega - \alpha_j(t)]}{\prod_{j=1}^{m} [\omega - p_j(t)]} . \tag{2.21}$$

Proceeding along the lines developed by Shraiman and Bensimon (1984), after some algebraic manipulations one can obtain the "pole dynamics" equations: a system of m ordinary differential equations governing the motion of the critical points of the map, $\alpha_j(t)$, of the form:

$$\partial_t \alpha_j = F_j(\alpha, p) , \tag{2.22a}$$

$$\partial_t p_j = G_j(\alpha, p) . \tag{2.22b}$$

Note, that while the p_j's completely determine the α_j's and vice versa (from Eq. (2.19–2.21) since the relation involves high order algebraic equations), it is more convenient to track the evolution of zeroes and poles by differential equations.

The simplest example is the case in which there is only one term in the sum in Eq. (2.19)

$$f_t(\Phi) = a_0(t) + \Phi/U + \frac{W}{2\pi} \ln (\omega - a_1(t)) . \tag{2.23}$$

Equations (2.22) then have the solution

$$a_0(t) = 2Ut ,$$

$$a_1^2(t) = 1 + [a_1^2(0) - 1] \, e^{-2\pi t U/W} . \tag{2.24}$$

In this case a cusp does not appear and instead the solution asymptotically approaches the Saffman-Taylor $\lambda = \frac{1}{2}$ steady-state solution.

The existence of a "pole" decomposition is somewhat surprising, since its existence is more commonly associated with integrable systems such as the

Burgers equation (Calogero, 1975; Chudnovski, 1977), and the KdV equation, (Kruskal, 1974; Moser, 1975), although it has been discovered in few other systems as well, see Lee and Chen (1982) and Thual, Frish and Henon (1985).

The differential equations (Eqs. 2.22) can be solved explicitly in some cases, otherwise they can be studied numerically. It can be shown (Sarkar, 1984, Howison, 1985a), that most initial conditions of the form (2.21), lead to the appearance of $\frac{2}{3}$ power cusps on the interface which appear when one of the zeroes of $\partial_\Phi f$ hits the unit disk. Figure 2.1 shows an example of the appearance of such a cusp in the evolution of the initial interface given by

$$f(\Phi) = \Phi/U + \sum_1^4 \log(\omega - p_j(t)) + C_0(t) , \tag{2.25}$$

with $p = (3, \ 10i, \ -9i, \ -6 + 4i)$. For most initial conditions, cusps form and after they form the equations seem to stop having solutions.

Physically, one expects the surface tension to prevent the singularities from appearing. Its effect should become important when the curvature near the cusp becomes of the order of the capillary length $W\sqrt{d_0}$. It may be possible to understand this problem using the method of matched asymptotic expansions (see Bender and Orzag, 1978), however, this has not yet been done.

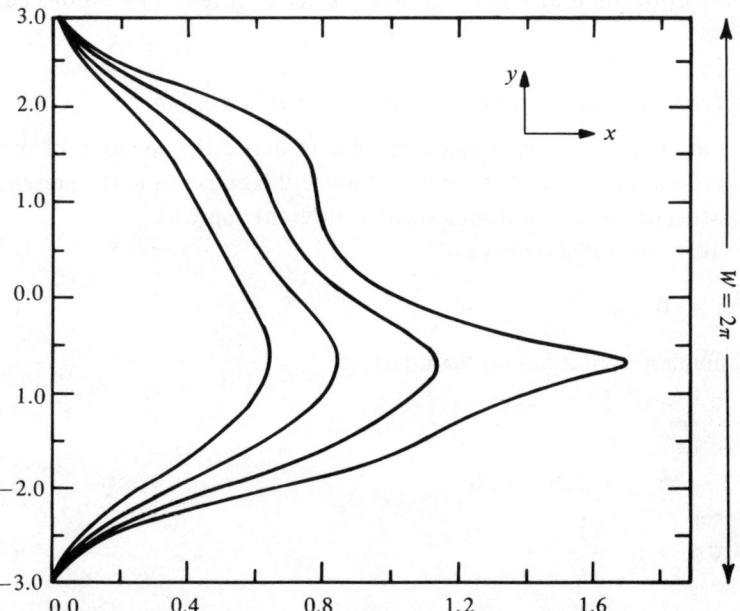

Fig. 2.1. Generation of a finite time cusp in the evolution of an arbitrary initial interface in the absence of surface tension.

3. The Conformal Mapping Algorithm

3.1. Introduction

There are several different ways of studying the fluid flow which results when the surface tension is not zero. Some of them have already been mentioned. Tryggvason and Aref (1983 and 1985) have applied a method in which the interface is described as a vortex sheet and thereby constructed a set of very appealing simulations of time development. A slightly different boundary integral method was employed by DeGregoria and Schwartz (1985) in their simulations of interface evolution. Both groups find that apparently stable fingers are generated, at least for the not-too-small values of d_0 for which their calculations are accurate. Their fingers, in turn, look very much like the fingers obtained from a direct solution of the steady-state problem as set up by McLean and Saffman (1981).

In this section we shall describe yet another method. This one is based upon the conformal mapping approach of Shraiman and Bensimon (1984) and used in the simulations carried out by Bensimon (1985). Similar ideas were used by Menikoff and Zemach (1983) in simulating motion of interfaces. Our motivation for this particular focus is that this method seems particularly well suited to the study of the limit of small d_0 where the behavior remains a bit of a puzzle.

3.2. Conformal method for the problem with surface tension

In Sec. 2, the hodograph technique was used to derive the equation of motion of the interface. In this section we will rederive this equation in the presence of surface tension ($d_0 \neq 0$) and from a slightly different approach.

The velocity potential ϕ obeys

$$\nabla^2 \phi = 0 \ , \tag{3.1a}$$

and the boundary conditions on the interface

$$\phi = \frac{Tb^2}{12\mu} \kappa \ , \tag{3.1b}$$

$$\hat{n} \cdot \nabla \phi = \hat{n} \cdot \frac{\partial \gamma}{\partial t} \ , \tag{3.1c}$$

where γ is the interface between the two fluids, κ its curvature and $\hat{n}$ indicates the direction normal to the interface. Instead of including boundary conditions at the side walls, we assume here a periodicity under $y \rightarrow y + W$. This corresponds physically to a cylindrical Hele-Shaw cell such as the one used by Aribert (1970).

Equations (3.1) determine the evolution of the interface: the Laplace equation with the Dirichlet boundary condition on γ, Eq. (3.1b), completely determines the flow field, then, the value of the normal velocity (the normal gradient of ϕ) at the boundary determines the velocity of the interface, Eq. (3.1c). We have to solve a Stefan, or, moving boundary value problem. The two-dimensionality of the problem greatly simplifies the task by allowing the use of the conformal mapping technique.

The idea which is standard in all textbooks on complex variables, e.g., Carrier, Krook and Pearson (1966) is based on the Riemann mapping theorem. This theorem ensures the existence of a conformal map from the complicated but simply connected domain enclosed by the interface γ into a standard domain, the interior of the unit disk. Within the disk the Dirichlet problem for the potential ϕ, Eq. (3.1a, b) can be readily solved. That solution then enables us to rewrite Eq. (3.1c) as an evolution equation for the mapping. As in Sec. 3, we introduce the complex potential $\Phi(z) = \phi(x,y) + \psi(x,y)$ (with $z = x + iy$). We then conformally map the domain of interest, i.e., the space occupied by the driven fluid, into the unit disk ($|\omega| \leqslant 1$: $z = f_t(\omega)$) (see Fig. 3.1). Since the interface γ between the two fluids in the image of the unit circle ($|\omega| = 1$) under the map $f_t(\omega)$:

$$\gamma(t, s) = f_t(e^{is}) , \qquad (3.2)$$

specifying the mapping, $f_t(\omega)$, at a given time t, is identical to specifying the interface, $\gamma(t, s)$ together with its parametrization, s.

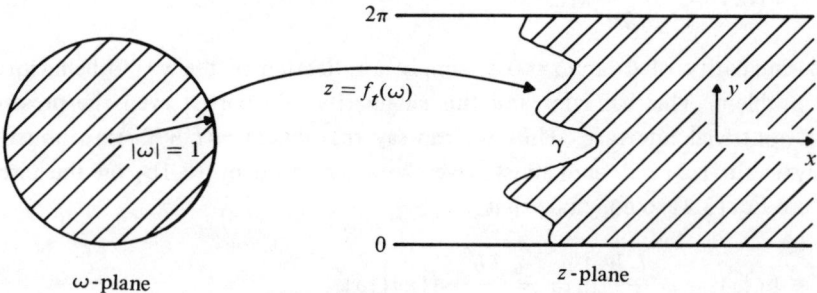

Fig. 3.1. The conformal map from the space occupied by the driven fluid to the unit disk.

In the unit disk the solution of the Dirichlet problem is standard. One has to find the function analytic inside the unit disk, $\Phi(\omega)$, real part which on the boundary, $\omega = e^{is}$, is specified, $\phi(s) = (Tb^2/12\mu)\kappa(s)$, where $\kappa(s)$, the local curvature of $\gamma(t, s)$ is

$$\kappa(s) = -\operatorname{Im} \frac{\partial_s^2 f / \partial_s f}{|\partial_s f|} . \tag{3.3}$$

The solution is known to be given by the Poisson integral formula. That formula states that the function analytic for $|\omega| < 1$, $\tilde{g}(\omega) = g(\omega) + ih(\omega)$, for which the real part on the unit disk $g(s)$ can be written as

$$g(s) = a_0 + \sum_{n=0}^{\infty} (a_n e^{ins} + a_n^* e^{-ins}) \tag{3.4a}$$

must be

$$\tilde{g}(\omega) \equiv A\{g\}(\omega) = a_0 + 2 \sum_{n=0}^{\infty} a_n \omega^n . \tag{3.4b}$$

Here we interpret $A\{g\}$ as an integral operator applied to a real-valued function defined on the unit circle, giving a new complex-valued function defined and analytic within that circle.

This approach can be directly applied to the determination of the potential. First we do the trivial case in which we have a flat interface and the curvature vanishes. In this case the potential is $\Phi = Uz$, while the mapping that takes one from the unit disk $|\omega| \leqslant 1$ onto the strip $x > 0$; $-W/2 \leqslant y \leqslant W/2$ is $z = -(W/2\pi)\ln \omega$. Hence in this example

$$\Phi(\omega) = -\frac{UW}{2\pi} \ln \omega . \tag{3.5}$$

The singularity of Φ at $\omega = 0$ is simply a reflection of the $x \to \infty$ behavior of the problem. This behavior and this singularity will persist even the presence of a nontrivial interface. Thus we can say that $\Phi(\omega) + (UW/2\pi)\ln \omega$ must be analytic for $|\omega| \leqslant 1$, and must have the value given by (3.1b) on the circle. We use Eq. (3.4) to conclude that

$$\Phi(\omega) = -\frac{UW}{2\pi} \ln \omega + \frac{Tb^2}{12\mu} A\{\kappa\}(\omega) .$$

Following the notation of the theorem, we use the notation $\tilde{\kappa}(\omega)$ for $A\{\kappa\}(\omega)$ and hence get an expression for the complex potential

$$\Phi(\omega) = -\frac{UW}{2\pi}\ln\omega + \frac{Tb^2}{12\mu}\,\tilde{\kappa}(\omega)\ . \tag{3.6}$$

A comparison between Eq. (2.20) and (3.6) will show that the ω used here will reduce to the ω of Sec. 2 as $T \to 0$. However the s's of the two sections are different. As $T \to 0$, the s of this section is $-2\pi/UW$ times the s of the previous section. Now, the normal velocity of the interface, given by Eq. (3.1c), is $(\hat{n} \cdot \nabla)\phi$ where $\hat{n}$ is a unit vector normal to the interface. This vector can be rewritten in complex notation as

$$n = n_x + in_y = i\,\frac{\partial_s f}{|\partial_s f|} = -\frac{\omega\partial_\omega f}{|\omega\partial_\omega f|}\ . \tag{3.7a}$$

Here and in the rest of this section, ω is specified to lie on the unit circle. Given n one can calculate the normal component of the gradient as

$$(\hat{n} \cdot \nabla)\phi = \partial_x \phi n_x + \partial_y \phi n_y = \partial_x \phi n_x - \partial_x \phi n_y$$

$$= \text{Re}\,(n\partial_z \Phi) = \text{Re}\left(n\,\frac{\partial_\omega \Phi}{\partial_\omega z}\right)$$

$$= -\frac{\text{Re}\,(\omega\partial_\omega \Phi)}{|\omega\partial_\omega f|}\ . \tag{3.7b}$$

Notice that Eq. (3.1c) only specifies the normal velocity of the interface. Of course there is no physical significance to a tangential velocity which simply corresponds to a reparametrization of the interface.

However the analyticity of the mapping function $f(\omega)$ fixes a particular "analytic," parameterization "gauge." This parametrization has to be maintained for all t. For that purpose, it is sufficient to make the time derivative of the map, $\partial_t f$, analytic inside the unit disc. To achieve this as in Eq. (2.8), we add to the right-hand side of Eq. (3.6) an appropriate tangential velocity component.

$$\partial_t f = n(\hat{n} \cdot \nabla)\phi + inC'$$

$$= \omega\partial_\omega f\left\{\frac{\text{Re}\,(\omega\partial\omega\Phi)}{|\omega\partial\omega f|^2} + iC\right\}\ . \tag{3.8}$$

Here C and C' are real functions of ω. To make the right-hand side of Eq. (3.9) analytic, the function C has to be the harmonic conjugate of the first term in the brackets $\mathrm{Re}(\omega\partial_\omega\Phi)/|\omega\partial_\omega f|^2$. In other words the terms in the brackets have to represent the function analytic in $|\omega|\leqslant 1$ and which real part on $|\omega|=1$ is specified. We have seen previously that this is achieved by the Poisson integral formula as expressed in Eq. (3.4). Therefore using Eq. (3.4) upon (3.8) and then substituting from Eq. (3.6) yields the desired evolution equation for the interface.

$$\frac{\partial f}{\partial t} = -\omega\partial_\omega fA\left\{\frac{1-(d_0W/2\pi)\,\mathrm{Re}(\omega\partial_\omega\tilde{\kappa}(\omega))}{|\omega\partial_\omega f|^2}\right\}\frac{UW}{2\pi}\,. \tag{3.9}$$

3.3. Numerical simulations

If we know $f_t(\omega)$ for a given value of t we can obtain the entire right-hand side of Eq. (3.9) and thus find $\partial_t f$. This enables one to set up a numerical algorithm for simulating the evolution of $\gamma(s,t)$. This algorithm is rather efficient. One measure of the quality is the number of operations needed per time step. In this case, if one fits the interface at N points and thus retains N coefficients in a Fourier series like (3.4), then the computer code requires $O(N\log N)$ operations per time step. (Most of the time is spent in computing Fourier series, using a Fast Fourier Transform algorithm.)

The algorithm was checked against known results in the asymptotic regime $(t\to\infty)$. A typical outcome of such a simulation fits well the finger shape obtained by the phenomenological scaling hypothesis of Pitts (1980), see Figs. 1.5 and 3.2b. The dependence of the finger width on the McLean-Saffman surface tension parameter κ is shown in Fig. 1.4 and agrees with their numerical results for the steady-state interface. In the absence of surface tension, the time evolution of the interface was in complete agreement with the exact time dependent solution. It developed finite time singularities, see Figs. 3.1 and 3.2a. In its presence one observes two regimes. One is at low velocities $(d_0 > 10^{-2})$ for which an initial arbitrary interface evolves into the corresponding McLean-Saffman steady-state propagating finger, Fig. 3.2b. The other is at high velocities $(d_0 < 10^{-2})$ for which the finger is unstable, and shows wobbling and tip splitting, Fig. 3.2c. This is in qualitative agreement with recent numerical and experimental work, Tabeling and Libchaber (1985), Park and Holmsy (1985), Liang (1985), DeGregoria and Schwartz (1985).

In Sec. 4 we will use this algorithm to study the linear and nonlinear stability of the propagating finger.

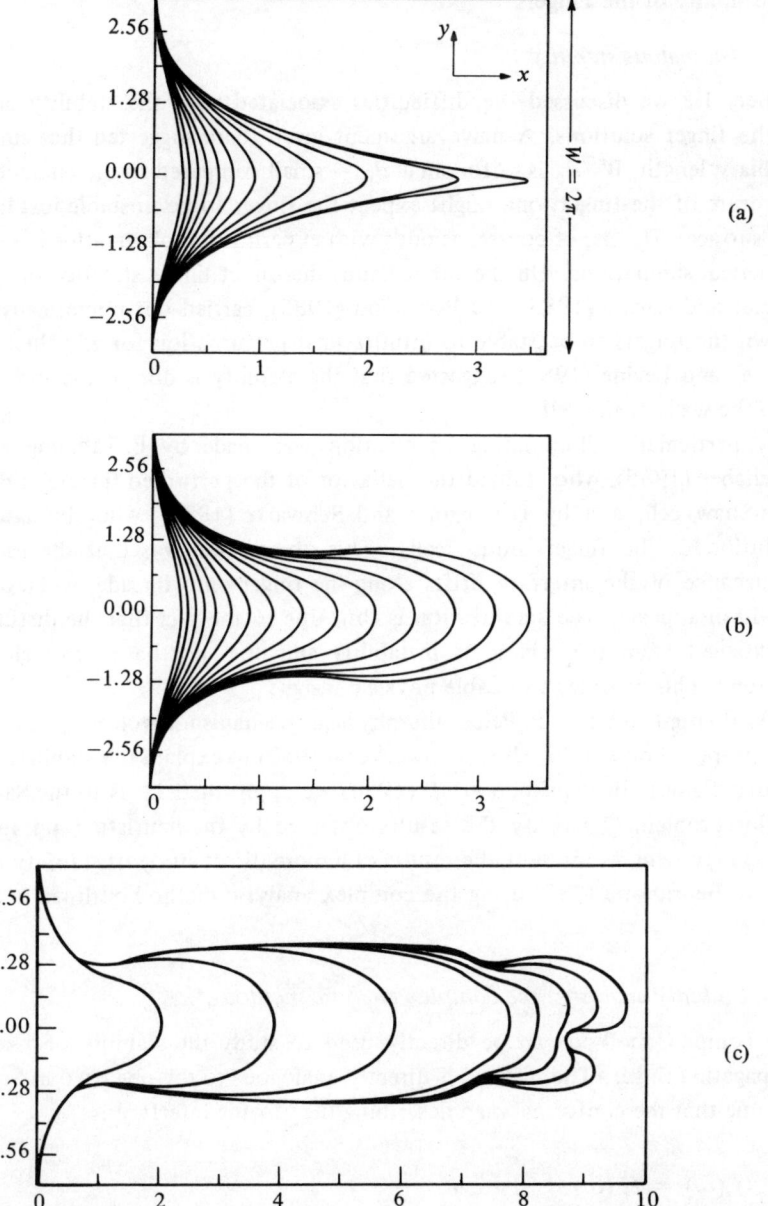

Fig. 3.2. Time evolution of an interface, from Bensimon (1985). Here the cell width W is set equal to 2π. (a) Evolution of an arbitrary initial interface without surface tension. (b) Evolution of the same initial interface as in (a), but in the presence of surface tension ($d_0 = .01$). (c) Tip splitting in the evolution of an interface at low surface tension ($d_0 = .01$).

4. Stability of the Fingers

4.1. Anomalous stability

In Sec. 1.3 we discussed the difficulties associated with the stability analysis of the finger solutions. A naive argument given there suggested that since the capillary length, $W\sqrt{d_0}$ is — for small d_0 — small compared to the characteristic curvature of the finger, one might expect the finger to be unstable just like the flat surface. This is, of course, at odds with experimental observations as well as numerical simulations. On the other hand, the direct linear stability analysis of Kessler and Levine (1985) and Bensimon (1985), carried out numerically, have shown the fingers to be stable to infinitesimal perturbation for all values of d_0. Kessler and Levine (1985) suggested that the stability is due to the interaction with the walls of the cell.

A particularly illuminating observation was made by P. Tabeling and A. Libchaber (1985), who studied the behavior of the perturbed fingers in the real Hele-Shaw cell, and by DeGregoria and Schwartz (1985), who simulated the evolution of the fingers numerically. This observation was that the localized disturbance of the interface drifts along the finger onto its side, whereupon it slowly disappears. The stabilization is thus due to the fact that the disturbance is expelled from the region of instability and does not have "enough time" to grow! This provides a valuable physical insight.

As pointed to us by P. Pelce, the physical mechanism involved is exactly the one proposed by Zel'dovich and coworkers (1980) to explain the stabilization of cellular flames. In Bensimon *et al.* (1986) we apply their ideas to the Saffman-Taylor problem. To verify the results obtained by the heuristic (and approximate) argument we present the results of a more direct study of stability carried out by Bensimon (1985) using the complex analytic method outlined in Sec. 3.

4.2. Stability analysis: the complex analytic method

The complex method can be directly used to study the stability of a steadily propagating finger. The method is directly analogous to the one used in Sec. 1.2. Assume that the conformal map describing the moving interface is

$$f_t(\omega) = f^0(\omega) + A_t(\omega) \ . \tag{4.1}$$

Here $f^0(\omega)$ is the steady state finger solution, which then depends upon d_0, and $A_t(\omega)$ is a small time-dependent deviation from this solution. Since $f_t(\omega)$ is analytic inside the unit disk we may assume

$$f^0(\omega) = \sum_{n=0}^{\infty} f_n \omega^n ,$$

$$A_t(\omega) = \sum_{n=0}^{\infty} A_n(t) \omega^n . \tag{4.2}$$

The stability analysis is done by expanding Eq. (3.9) and keeping terms of first order in A. Since f^0 is a steady-state solution, the result is an equation of the form

$$\dot{A}_k(t) = \sum_{n=1}^{\infty} M_{kn}[f^0] A_n(t) + M'_{kn}[f^0] A_n^*(t) . \tag{4.3}$$

Notice that the matrices M and M' depend upon the presumed steady-state solution f^0. The $k=0$ term in Eq. (4.3) gives $\partial_t A_0 = 0$, thus yielding a marginal mode that corresponds to the translation of the finger. The fact that it decouples from the rest is quite convenient and is an advantage of the method.

The solution to the linear stability problem can now be clearly seen. Consider M and M' in Eq. (4.3) to be matrices. Form the supermatrix

$$M = \begin{bmatrix} M & M' \\ M'^* & M^* \end{bmatrix} . \tag{4.4}$$

If that matrix has an eigenvalue E, then $A_n(t)$ is exponential in t, e^{Et}. Stability then requires that the real part of the eigenvalue be negative so that any deviation from the finger solution would vanish as $t \to \infty$.

Once the the matrix elements are written down, the eigenvalues can be calculated numerically. Bensimon (1985) expanded f^0 in a power series in d_0 and then used that expansion to calculate M and M' to first order in d_0. The eigen-spectrum for $d_0 = 0.05$ is shown in Fig. 4.1. Notice the continuum of of symmetric modes (A_n real) and antisymmetric modes (A_n imaginary) with a negative real eigenvalue preceded by a discrete set of asymmetric modes with complex eigenvalues (their number increases as $d_0 \to 0$). The eigenvalues all have negative real part for values of d_0 down to 10^{-3}, so that the interface appears to be linearly stable. This is in agreement with the heuristic argument described in Bensimon *et al.* This stability result, as well as the eigenvalue spectrum, is also in agreement with previous results of Kessler and Levine (1985). It disagrees with the results of McLean and Saffman (1981), where instability was predicted. However Sarkar has argued that the instability prediction might be wrong because of neglected terms of order d_0 while Levine suggested that

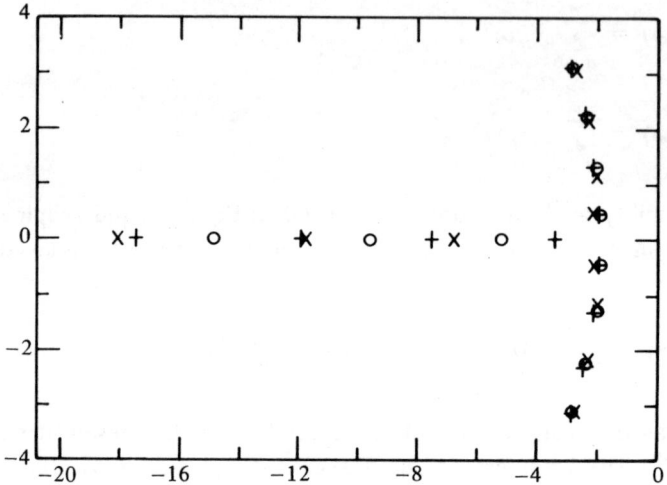

Fig. 4.1. The eigenspectrum at $d_0 = .05$, for various truncations: $N = 130$ (+); $N = 80$ ($\times$); $N = 100$ ($\circ$). Notice the discrete spectrum of asymmetric modes (complex eigenvalues) and the continuum of symmetric and antisymmetric modes (negative real eigenvalues), from Bensimon (1985).

the number of mesh points was not sufficient to ensure accuracy. It appears to us that the results of Kessler and Levine (1985) and Bensimon (1985) are more reliable.

4.3. Structural stability and nonlinear instability

We have previously seen that the finger is destabilized due to the existence of a finite amplitude instability. Its proximity shows up as a sensitivity of the eigenspectrum of the linearized problem. Bensimon (1985) argued, that there is a relation between the structural stability of the linearized problem and the nonlinear instability of the full problem.

One may then study the dependence of the critical amplitude for destabilization on d_0 by looking at the appearance of unstable modes in response to a random distortion of the interface (letting the f_m in Eq. (4.2) contain a random term), an instability arising at a typical perturbation strength v_c which depends upon d_0. For example, we can obtain a fit with a form

$$v_c \sim d_0^{1/2} \exp(-\beta d_0^{-1/2}) , \qquad (4.5)$$

with $\beta \approx 1.3$. This particular form for the fit was suggested by the argument of

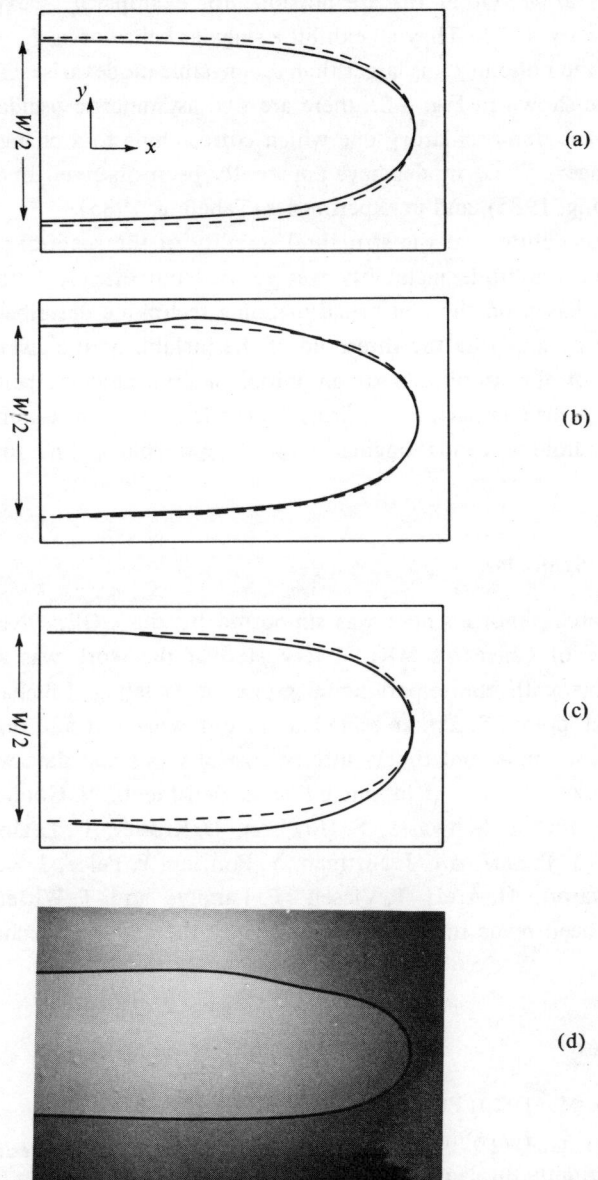

Fig. 4.2 The three most unstable modes (dashed line: unperturbed finger). (a) The symmetric non-oscillatory mode, which may be responsible for the experimentally and numerically observed fingers of width $\lambda < 1/2$. (b) The asymmetric "hump" mode. (c) The asymmetric "tip wobbling" mode. (d) The asymmetric "hump" mode observed in an experiment (courtesy of P. Tabeling and A. Libchaber (1985)).

Bensimon *et al.* Other fits are possible, for example $v_c \approx \exp(-\gamma d_0^{-\beta})$, with $\beta \approx .61$ and $\gamma \approx .72$. They all exhibit a singular behavior as $d_0 \to 0$.

When the noise in f^0 is larger than v_c unstable modes arise. The most unstable modes are shown in Fig. 4.2: there are two asymmetric oscillatory modes, and a symmetric non-oscillatory one which corresponds to a change in the width λ of the finger. These modes have apparently been observed in numerical simulations (Liang, 1985), and in experiments (Tabeling, 1985).

The conjecture that the structural stability of the spectral problem is related to a finite amplitude instability was verified numerically with the help of the algorithm based on the conformal mapping technique described in Sec. 3. This was done by studying the threshold of the instability at a given value of d_0 as a function of the amplitude of an initial random analytic perturbation of the interface. This comparison confirms the existence of a finite amplitude instability with threshold depends singularly on d_0, possible of the form predicted by Eq. (4.5).

Acknowledgments

The research reported here was supported by the DOE, ONR, NSF, and the University of Chicago's MRL. Every step of the work was aided by a lively interchange with the experimental group of Tabeling, Libchaber, and Heslot. At several points S. Sarkar joined us in our work. In addition we have been helped by a frank and timely interchange of views and data with many of the other workers in the field, including E. Ben-Jacob, N. Goldenfeld, J. Langer, A. DeGregoria, L. Schwartz, S. Howison, D. Kessler, H. Levine, J. Koplik, R. Lenormand, P. Saffman, J. Nittman, Y. Pomeau, P. Pelce, J. Kertesz, U. Frisch, G. Tryggvason, H. Aref, T. Vicsek, F. Family, and T. Witten. Work in this area has been made much more pleasant by the vigorous exchanges mentioned above.

References

Aribert, J.-M. (1970) PhD thesis, Toulouse.

Barenblatt, G. I. (1979) *Similarity, Self-similarity and Intermediate Asymptotics* (Consultants Bureau, New York).

Barenblatt, G. I., Zel'dovich, Ya. B. (1972) *Ann. Rev. Fluid Mech.* **4**, 285.

Bender C. M., Orsag, S. A. (1978) *Advanced Mathematical Methods for Scientists and Engineers* (McGraw-Hill, New York).

Ben-Jacob, E., Goldenfeld, N., Kotliar, G., Langer, J. (1984) *Phys. Rev. Lett.* **53**, 2110.

Bensimon, D. (1986) *Phys. Rev.* **A33**, 1302.

Bensimon, D., Kadanoff, L. P., Liang, S., Shraiman, B. I. and Tang, C. (1986) submitted to *Rev. Mod. Phys.*

Birkhoff, G., Zarantonello, E. (1957) *Jets, Wakes and Cavities*, New York.

Calogero, F. (1975) *Lett. Nuovo Cimento* **13**, 411.

Carrier, G. G., Krook, M. and Pearson, C. E. (1966) *Functions of Complex Variable* (McGraw-Hill, New York).

Chudnovsky, D. V. and Chudnovsky, G. V. (1977) *Nuovo Cimento* **40B**, 339.

Chuoke, R. L., van Meurs, P. and van der Pol, C. (1959) *Trans. AIME* **216**, 188.

DeGregoria A. J. and Schwartz, L. W. (1985) Exxon preprints.

DiBenedetto, E. and Friedman, A. (1984) *T. Am. Math. Soc.* **282**, 183.

Elliot, C. M. and Janovsky, V. (1981) *Proc. Roy. Soc.* Edin. **88A**.

Hele-Shaw, H. J. S. (1898) *Nature* **58**, 34.

Howison, S. D. (1985a) *J. Appl. Math.*, to appear in *SIAM*.

Howison, S. D., Ockendon, T. R. and Lacey, A. A. (1985b) to appear in *QJMAM* 267.

Kessler, D. A. and Levine, H. (1985) preprint.

Kessler, D. A., Koplik, J. and Levine, H. (1984) *Phys. Rev.* **A30**, 3161 and *Proceedings of the Electrochemical Soc. of Amer.*, Toronto (1985).

Kruskal, M. D. (1974) in *Nonlinear Wave Motion*, Newell, A. ed., (Am. Math. Soc., Prov. R. I.) p. 61.

Langer, J. S. (1985) ITP preprint.

Lee, Y. C. and Chen, H. H. (1982) *Phys. Scripta* **T2**, 41.

Liang, S. (1986) *Phys. Rev.* **A33**, 2663.

Maher, J. (1985) *Phys. Rev. Lett.* **54**, 1498.

McLean, J. W. and Saffman, P. G. (1981) *J. Fluid Mech.* **102**, 455, and McLean, J. W. (1980) PhD thesis, Caltech.

Menikoff, R., Zemach, C. (1983) *J. Comp. Phys.* **51**, 28.

Meyer, G. H. (1982) in *Numerical Treatment of Free Boundary Value Problem*, J. Albercht, ed., (Birkhauser, Basel).

Moser, J. (1975) *Adv. Math.,* **16**, 197.

Nittman, J., Daccord, G. and Stanley, H. E. (1985) *Nature* **314**, 141.

Park, C.-W. and Homsy, G. M. (1984) *J. Fluid Mech.* **139**, 291.

Park, C.-W. and Homsy, G. M. (1985) *Phys. Fluids* **28**, 1583, 1621.

Pietronero, L. and Wiesmann, H. J. (1984) *J. Stat. Phys.* **36**, 909.

Pitts, E. (1980) *J. Fluid Mech.* **97**, 53.

Pomeau, Y. and Pelce, P. (1985) preprint.

Richardson, S. (1972) *J. Fluid Mech.* **56**, 609.

Saffman, P. G. and Taylor, G. I. (1958) *Proc. R. Soc.* Lond. **A245**, 312.

Sarkar, S. (1984) *Phys. Rev.* **A31**, 3468.

Shraiman, B. I. and Bensimon, D. (1984) *Phys. Rev.* **A30**, 2840.

Tabeling, P. and Libchaber, A. (1986) *Phys. Rev.* **A33**, 794.

Thual, O., Frish, U. and Henon, M. (1985) Nice Observatory preprint.

Tryggvason, G. and Aref, H. (1983) *J. Fluid Mech.* **136**, 1, and to be published, 1985, *J. Fluid Mech.*

Vanden-Broeck, J.-M. (1983) *Phys. Fluids* **26**, 2033.

Zel'dovich, Ya. B., Istratov, A. G., Kidin, N. I. and Librovitch, V. B. (1980) *Comb. Sci. and Tech.*, **24**, 1.

Note added in proof:

The reader may also be interested in more recent papers:

Combescot, R., Dombre, T., Hakim, V., Pomeau, Y. and Pumir, A. (1986) *Phys. Rev. Lett.* **56**, 2036.

Hong, D. C. and Langer, J. S. (1986) *Phys. Rev. Lett.* **56**, 2032.

Shraiman, B. I. (1986) *Phys. Rev. Lett.* **56**, 2028.

PATTERN RECOGNITION BY FLEXIBLE COILS

P. G. de Gennes

College de France
75231 Paris Cedex 05
FRANCE

We discuss a two-stage process: (a) *writing* — one conformation $(\mathbf{r}_1, \mathbf{r}_2, \ldots, \mathbf{r}_N)$ of a gaussian chain (generated statistically, at a temperature T_0) is printed into a d-dimensional substrate; (b) *reading* — the coil (at a new temperature T_1) is put into contact with the print; a strong, non specific, attraction is imposed between the coil and the print: each coil site must stick to a print site, but not necessarily in the right order. A formal rule, giving the weight of any specific permutation P in the reading sequence, is written down. In a certain approximate limit the problem reduces to a discussion of self avoiding Levy flights along the chemical sequence. Then we find that for $d > 4$ the image is faithful, and corresponds to sequential reading of the print, except for local noise effects. But, for $2 \leqslant d < 4$ the sequence is wrong. However, for all $d > 2$ the image remains significant, i.e., the shape of the coarse grained print is reproduced. At $d = 2$ the pattern is completely blurred.

1. Introduction

Pattern recognition by macromolecules is a classical feature of living systems e.g., in immunology, or with transfer RNAs, or (at a more complex level) in protein renaturation — a process which can be loosely described as self recognition (Westlaufer, 1984). In most biological situations the relevant peptide, or nucleotide chains build up structures which are partly rigid; also some relevant functional groups have strong stereospecific interactions.

Our aim in the present article is to define an oversimplified version of these problems, based on linear flexible, ideal coils. However, if, in some (remote) future, information storage based on synthetic macromolecules becomes a reality, the ideal coil system may provide a useful limiting case.

1.1. Writing

The general idea is to start with a flexible template carrying a certain number of identical units $(1, 2, \ldots, N)$ and to have one particular conformation of the template $(\mathbf{r}_1, \mathbf{r}_2, \ldots, \mathbf{r}_N)$ "printed" on a certain substrate. In two space dimensions $(d = 2)$ this operation is easy to visualise. One can, for instance, think of the sites $(1, \ldots, N)$ as catalytic sites. When the template lies on a flat, reactive, surface, each catalytic site induces a certain local transformation on this surface. We can remove the template, but keep a "print" on the plane. In 3 dimensions $(d = 3)$ the process is less familiar, but one can think of a chain penetrating a swollen gel, and again transforming it locally to produce a print. As usual in statistical mechanics, we shall find it useful to work at arbitrary dimensionalities d (up to $d = 4$ or more in the present case).

The statistical weight for a template conformation $(\mathbf{r}_1, \ldots, \mathbf{r}_N)$ has the standard Gaussian form

$$W_0(\mathbf{r}_1, \ldots, \mathbf{r}_N) = \prod_{i=1}^{N-1} \left(\frac{\beta_0}{2\pi s_i}\right)^{d/2} \exp\left(-\frac{\beta_0 a_i^2}{2 s_i}\right) , \qquad (1.1)$$

where

$$\mathbf{a}_i = \mathbf{r}_{i+1} - \mathbf{r}_i ,$$

$$s_i = m_{i+1} - m_i . \qquad (1.2)$$

Here $i = 1, \ldots, N$ labels the units, and s_i is a measure of the chemical distance between adjacent units. In most of this work, we shall consider only equidistant units (s_i = constant) and, by suitable normalisation we shall take $s_i \equiv 1$.

1.2. Reading

Having generated a "print", we want to understand how it can be recognized by another coil, chemically identical to the first. We shall allow however for certain differences (in temperature, pH, etc) between the reading conditions and the writing conditions: for instance, the persistence length l of the chain (Birshtein and Ptitsyn, 1966) may change. The essential parameter describing this change is the ratio of the natural (mean square) end-to-end sizes of the chain. We call this ratio, x, the *mismatch parameter*:

$$x = \frac{R_1^2}{R_0^2} = \frac{l_1}{l_0} \quad , \tag{1.3}$$

where R_0 refers to writing, and R_1 to reading conditions.

At the reading stage, the chain $(1, \ldots, N)$ is subjected to a strong, non selective attraction from the printed sites $(\mathbf{r}_1, \ldots, \mathbf{r}_N)$. Each printed site traps one, and only one, chain site. But, the ordering may be different: the unit (i) sticks to a site $\mathbf{r}_{Pi}$ where Pi is the transform of i by a certain permutation P. The statistical weight for this conformation is

$$W_1(\mathbf{r}_1, \ldots, \mathbf{r}_N | P) = \prod_i \left(\frac{\beta_1}{2\pi s_i}\right)^{d/2} \exp\left(- \frac{\beta_1 a_{Pi}^2}{2s_i}\right) . \tag{1.4}$$

Here

$$a_{Pi} \equiv \mathbf{r}_{P(i+1)} - \mathbf{r}_{Pi} \quad .$$

The coefficient β_1 is related to the mismatch parameter x via

$$x = \frac{\beta_0}{\beta_1} \tag{1.5}$$

1.3. Qualitative features

In the limit of small x, the reading chain is strongly stretched between conse-cutive points, and becomes asymptotically a sequence of linear segments: the geometrical pattern is reminiscent of a traveling salesman problem (Beardwood *et al.*, 1959; Kirkpatrick and Toulouse, 1985; Vannimenus and Mezard, 1985). In the opposite limit $(x \gg 1)$ each arc of the reading chain occupies a domain of linear size $\rho_1 \sim \beta_1^{-1/2}$ — much larger than the arc size of the print $\rho_0 \sim \beta_0^{-1/2}$ $\sim \rho_1 x^{-1/2}$. In all cases, there is a certain *coarse-grained tube*, of diameter

$$\rho^2 = \rho_0^2 - \rho_1^2 \tag{1.6}$$

which surrounds the original print, and which contains the reading image (Fig. 1). The tube volume Ω is of order N while the overall coil volume is $\sim R_0^d \sim N^{d/2}$. Thus, whenever $d > 2$, the ratio $\Omega/R_0^d \sim N^{1-d/2}$ is very small at large N: the image has a (coarse-grained) similarity to the original print.

   This does not mean, however, that the *ordering* of the points r_{Pi} is right: for instance, if we have a large loop (as in Fig. 2) where the tube intersects itself, the reading chain may well remain in the tube, but run along the loop in the wrong direction (we call this *non-sequential reading*).

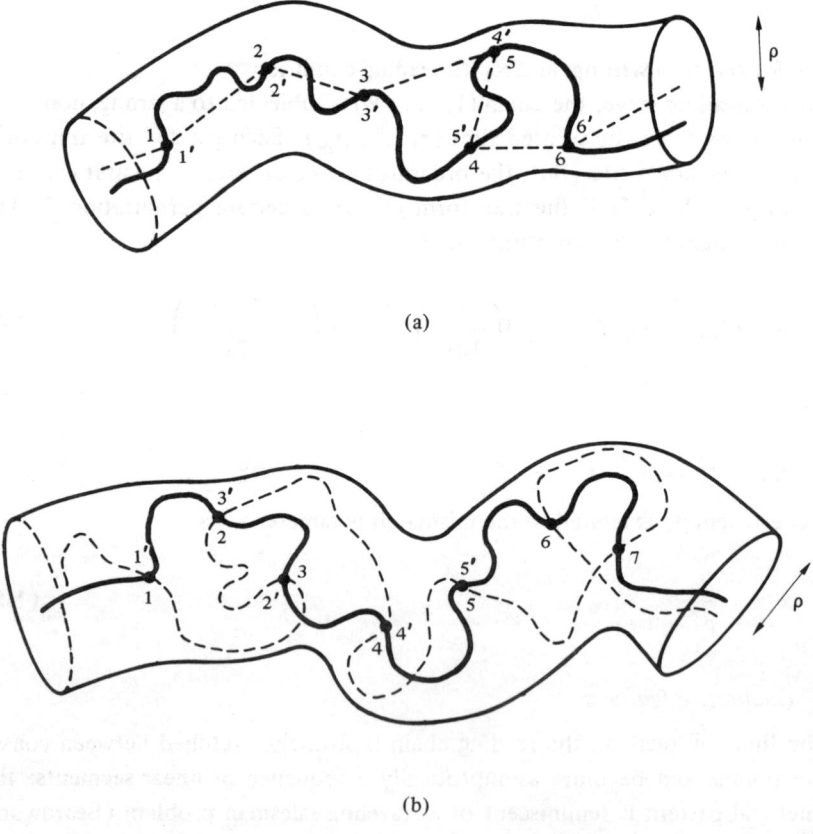

(a)

(b)

Fig. 1. The "print" (continuous line 1, 2, 3, ...) and the "reading chain" (dotted line 1', 2', 3', ...):
(a) a case in which the natural size of the reading chain would be smaller than that of the print ($x < 1$): the reading chain is strongly stretched;
(b) the reverse case ($x > 1$) with a higher density of mistakes in the reading sequence.

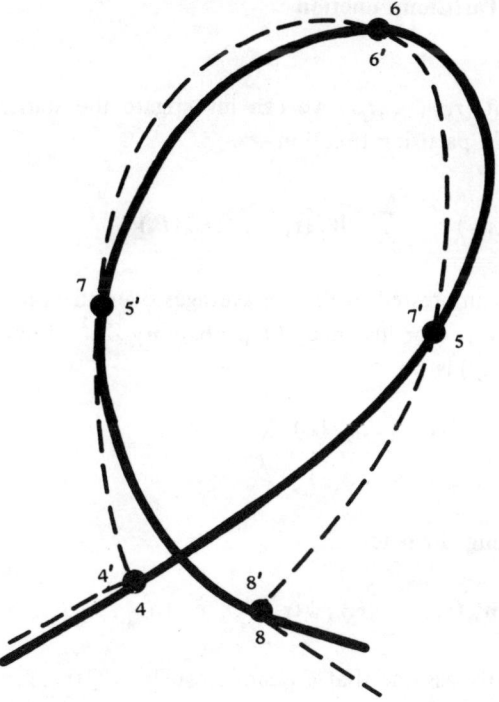

Fig. 2. A large loop in the print (continuous line 1, 2, . . .) can be followed in the wrong order by the reading chain (dotted line 1', 2', . . .), while keeping a high statistical weight.

Our main aim is to discuss this type of fault. The number of large loops in a chain is of order $N \Omega R_0^{-d} \sim N^{2-d/2}$ and is thus small for $d > 4$. Indeed, we shall find that $d = 4$ is the critical dimensionality for sequential reading. In Sec. 2, we construct the basic formula for the partition function $\langle \mathscr{Z}_N \rangle$ of the reading chains, averaged over all print conformations. We express $\langle \mathscr{Z}_N \rangle$ in terms of diagrams (each diagram describing a permutation P). In Sec. 3 we try to simplify the diagrammatic rules, and propose an approximate version, which leads as to a problem of Levy flights along the chemical sequence. Our flights are self avoiding: we incorporate this feature in Sec. 4. Also, our flights are "compressed" because of the requirement that all printing sites be filled by reading sites. The effects of this compression are presented in Sec. 5; the main tool here is to relate the quality of sequential reading to a certain entropy loss S due to the constraints. The results are summarised and criticized in Sec. 6.

2. The Average Partition Function

2.1. Definitions

For a given print $(\mathbf{r}_1, \ldots, \mathbf{r}_N)$ we can investigate the statistics of the reading chain in terms of a partition function

$$Q(\mathbf{r}_1, \ldots, \mathbf{r}_N) = \sum_{P=1}^{N!} W_1(\mathbf{r}_1, \ldots, \mathbf{r}_N \,|\, P) \ . \tag{2.1}$$

We are in fact interested in certain averages over all prints, defined with the original weights W_0. For instance, the probability Z_N^{-1} of having an exact image $(P = 1;$ the identity) is

$$Z_N^{-1} = \left\langle \frac{W_1(\mathbf{r}_1, \ldots, \mathbf{r}_N \,|\, I)}{Q(\mathbf{r}_1, \ldots, \mathbf{r}_N)} \right\rangle \ , \tag{2.2}$$

where the averaging law reads

$$\langle \psi \rangle = \int W_0(\mathbf{r}_1, \ldots, \mathbf{r}_N)\, \psi(\mathbf{r}_1, \ldots, \mathbf{r}_N)\, d\mathbf{r}_1 \ldots d\mathbf{r}_N \ . \tag{2.3}$$

We shall constantly assume that a quantity such as $Q(\mathbf{r}_1, \ldots, \mathbf{r}_N)$ has a narrow distribution, so that we can simplify ratios of the type given in Eq. (2.2)

$$Z_N^{-1} \ \rightarrow \ \frac{\langle W_1(\mathbf{r}_1, \ldots, \mathbf{r}_N \,|\, I) \rangle}{\langle Q \rangle} \ . \tag{2.4}$$

The quantity Z_N defined in Eq. (2.4), is a reduced partition function, and contains most of the desired information.

2.2. Weight of a diagram

Each permutation P can be described by a one-dimensional diagram (Fig. 3) showing how the chemical sequence of the reading chain maps on the print sequence. The weight of the permutation P is, from Eqs. (1.1 – 1.3).

$$\langle W_1(P) \rangle = \left\langle \prod_j \left(\frac{\beta_1}{2\pi} \right)^{d/2} \exp\left(-\frac{\beta_1 a_{Pj}^2}{2} \right) \right\rangle \ , \tag{2.5}$$

where we now assume that all bonds in the template are equal in length $(s_i \equiv 1)$.

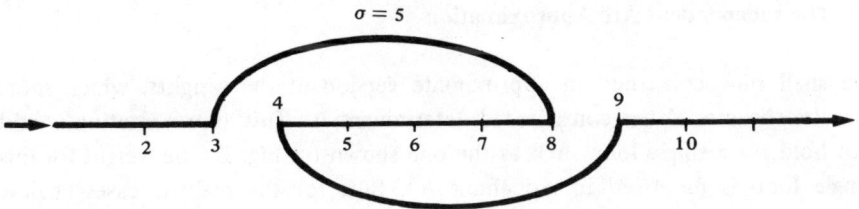

Fig. 3. A typical diagram associated with a permutation P. The particular case shown here corresponds to one relatively large loop read in the wrong order.

To eliminate some uninteresting prefactors, it is convenient to go to reduced weights

$$\widetilde{W}(P) \equiv \frac{\langle W_1(P) \rangle}{\langle W_1(I) \rangle} \ . \tag{2.6}$$

A natural method to compute Eq. (2.5) explicitly amounts to expressing the vectors a_{Pj} for each "arc" $(Pj, Pj+1)$, in terms of the original vectors, a_i. Then Eq. (2.5) is a multidimensional Gaussian integral, and can be expressed in terms of determinant, $\Delta(P)$

$$\widetilde{W}(P) = [\Delta(P)]^{-d/2} \ . \tag{2.7}$$

The rows and columns of Δ are associated with the bonds i (linking i and $i+1$ in the print chain). The diagonal elements are

$$\Delta_{ii} = \frac{n_i + x}{1 + x} \ , \tag{2.8}$$

where $n_i(P)$ is the number of arcs of the reading chain which "fly over" bond i of the printing chain, as read on the diagram associated with P. Similarly the off diagonal elements are

$$\Delta_{ij} = \frac{c_{ij}}{1 + x} \qquad (i \neq j) \ , \tag{2.9}$$

where c_{ij} is the number of arcs which "fly over" both i and j; again this number may be found by inspection of the diagram. Equations (2.7)–(2.9) give us a formal answer for the statistical weights. But there is still a long way to go to transform this into physical predictions.

3. The Independent Arc Approximation

We shall now construct an approximate version of the weights, which short circuits the use of the complicated determinant Δ. This approximation would not hold for a single loop such as the one shown on Fig. 2 (the weight for this single loop is discussed in Appendix A). But, for the realistic cases (below $d = 4$) where loops build up over loops, it may be a reasonable starting point.

Let us return to Eq. (2.5), and introduce first a set of vectors $\mathbf{k}_j$ conjugate to the arc vectors $\mathbf{a}_{Pj}$. Then

$$\langle W_1(P) \rangle = \left\langle \prod_j \int \frac{d\mathbf{k}_2}{(2\pi)} \, d \, \exp\left(i\mathbf{k}_j \cdot \mathbf{a}_{Pj} - \frac{k_j^2}{2\beta_1}\right) \right\rangle . \tag{3.1}$$

We now assume that the different arc vectors a_{Pi}, a_{Pj} can be treated as uncorrelated:

$$\left\langle \prod_j \exp(i\mathbf{k}_j \cdot \mathbf{a}_{Pj}) \right\rangle = \prod_j \langle \exp(i\mathbf{k}_j \cdot \mathbf{a}_{Pj}) \rangle . \tag{3.2}$$

Then, using a standard theorem on Gaussian averages, each factor is

$$\langle \exp(i\mathbf{k}_j \cdot \mathbf{a}_{Pj}) \rangle = \exp - \tfrac{1}{2} \langle (\mathbf{k}_j \cdot \mathbf{a}_{Pj})^2 \rangle$$

$$= \exp(-k_j^2 S_{Pj} / 2\beta_0) . \tag{3.3}$$

Now we can integrate (3.1) over the $\mathbf{k}$ variables, and obtain

$$\langle W_1(P) \rangle = \prod_j [2\pi(\beta_1^{-1} + \beta_0^{-1} S_{Pj})]^{-d/2} . \tag{3.4}$$

For the special case of the exact image ($P = 1$) we simply set $S_{Pj} = 1$ in (3.4). This now allows us to write the reduced weight in the form of a product over all arcs (of length σ):

$$\widetilde{W}(P) = \prod_{\text{arcs}} h_x(\sigma) , \tag{3.5}$$

$$h_x(\sigma) = \left(\frac{1+x}{|\sigma|+x}\right)^{d/2} . \tag{3.6}$$

With the rules (3.5, 6) we can think of each diagram as a self avoiding walk of N steps; but the jumps lengths for each step are not restricted to $\sigma = 1$: they have a distribution which (at large σ) is essentially a power law $h(\sigma) \sim \sigma^{-d/2}$. This limit corresponds to the classical Levy flights.

4. Self Avoiding Levy Flights

Self avoiding Levy flights have been analysed recently by Grassberger (1985), and Halley and Nakanishi (1985). Some central results will be briefly recalled below for the particular case of interest here (namely one-dimensional flight along the chemical sequence of a print).

4.1. A reminder on ideal Levy flights (Hughes et al., 1982)

The statistical weight for a flight of N steps, starting from the origin, and reaching a point m, will be called $G_N(m)$, and is ruled by a simple convolution equation:

$$G_{N+1}(m) = \sum_{\sigma \neq 0} G_N(m - \sigma) h_x(\sigma) \Big/ \sum_{\sigma' \neq 0} h_x(\sigma') \tag{4.1}$$

together with the initial condition $G_0(m) = \delta_{m0}$. Going to Fourier transforms

$$h_x(q) = \sum_{\sigma} h_x(\sigma) e^{iq\sigma} , \tag{4.2}$$

this gives the following structure:

$$G_N(q) \equiv \sum_{m} G_N(m) e^{iqm} = \exp(-N\omega(q)) , \tag{4.3}$$

where

$$\exp(-\omega(q)) = \frac{h_x(q)}{h_x(q = 0)} . \tag{4.4}$$

The limit of interest for us corresponds to large m or small q:

$$1 - \exp(-\omega(q)) \cong \omega(q) = \frac{\sum_{\sigma} h_x(\sigma)(1 - \cos q\sigma)}{\sum_{\sigma} h_x(\sigma)} . \tag{4.5}$$

Returning to the definition of $h(\sigma)$ (Eq. 2.11) we see that the sums converge strongly if $d > 6$, in which case $\omega(q) \sim q^2$. But for $d < 6$ the result is more singular

$$\omega(q) = k(x)q^{\mu} \quad,$$

$$\mu = d/2 - 1 \;, \tag{4.6}$$

where $k(x)$ is a prefactor. Inverting the Fourier transform, $G_N(m)$ has the scaling form

$$G_N(m) = M_0^{-1} f\left(\frac{m}{M_0}\right) \tag{4.7}$$

with

$$M_0 = N^{1/\mu} \;. \tag{4.8}$$

M_0 is the characteristic width of the distribution. For $d = 6$ we return to standard random walks with nearest neighbor jumps ($\mu = 2$) but whenever $d < 6, M_0$ is much larger than $N^{1/2}$.

4.2. Self avoiding flights

Each site on the chemical sequence is visited only once: in the language of Halley and Nakanishi (1985), this corresponds to a one-dimensional, "node avoiding" Levy flight. The width $M(N)$ of a flight of N jumps may now become larger than $M_0(N)$. There are three relevant regimes:

(a) for $\mu < \frac{1}{2}$ (or, equivalently, $d < 3$) the excluded volume effect is irrelevant, because the density ϕ of points achieved in the ideal flight ($\phi \sim N/M_0(M)$) is very small. The corresponding reduction of entropy is

$$S_{\text{ex}} = \frac{1}{2}N\phi \cong \frac{N^2}{M_0(N)} \cong N^{2-1/\mu} \tag{4.9}$$

and is negligible for $\mu < \frac{1}{2}$. Another (equivalent) presentation of this result is the following: Eq. (4.8) shows that the fractal dimension of the ideal flight is μ. The overlap of two flights in one dimension (or equivalently, the self overlap of one flight) becomes negligible when $2\mu < 1$.

(b) for $\mu > 1$ (or equivalently $d > 4$) we return to standard self avoiding walks. This may seem surprising at first sight, because the statistics of the ideal Levy flight are still very different from a conventional random walk (i.e., $M_0(N) > N^{\frac{1}{2}}$) for all the interval $1 < \mu < 2$ ($4 < d < 6$). We shall give here an intuitive argument which explains this in terms of a Flory (1971) approximation. The starting point is a free energy of the form

$$F(M) = F_{el} + F_{ex} \quad , \tag{4.10}$$

where the elastic term F_{el} is *not* proportional to M^2, as it is for conventional chains, but is (in reduced units)

$$F_{el} = \left(\frac{M}{M_0}\right)^{\mu/(\mu-1)} \quad (1 < \mu < 2) \quad . \tag{4.11}$$

A brief justification of Eq. (4.11) is given in Appendix B. The exclusion term F_{ex} has the standard form

$$F_{ex} = \frac{1}{2}N\phi \cong \frac{N^2}{M} \quad . \tag{4.12}$$

Minimizing the sum (4.10) we do arrive at $M = N$, i.e., to the conventional result for self-avoiding walks in one dimension.

(c) in the intermediate region $\frac{1}{2} < \mu < 1$ (or $3 < d < 4$) there is a non trivial swelling of the ideal flight due to excluded volume effects

$$M(N) \cong N^{\nu} \quad , \tag{4.13}$$

$$p^{-1} < \nu < 1 \quad . \tag{4.14}$$

Numerical values of $\nu(\mu)$ are discussed in Halley *et al.* (1986).

5. Confined Flights

The results summarized in Sec. 3 concern self-avoiding Levy flights of N steps on an *infinite* (one-dimensional) lattice. But our diagrams have a further constraint: we have stipulated that N consecutive sites of the print must be occupied by N sites of the reading chain. In terms of walks, this means that our walk is *confined* to a finite segment.

Consider, for instance, the case where the walk starts at one end of the confined segment (e.g. at $i = 0$) and progressively moves towards the other end ($i = N$). This amounts to assuming that the first reading site is imposed to be in the correct position. What happens to the others? Even for ideal Levy flights, this leads to a rather complex problem (of the Wiener-Hopf type) near the boundaries ($i = 0$ and $i = N$). Thus we shall find it convenient to define a slightly different problem. We make the print sequence *cyclic* (with N bonds) and ask how this cyclic structure can be explored by a flight of N steps.

(a) *For ideal flights* one can expand the statistical weight or propagator $G_N(m)$ in terms of eigenfunctions (here, Fourier component). The entropy reductions of a chain of N steps confined in a cycle of length L is then found to be

$$S(N, L) \cong \left(\frac{M_0(N)}{L} \right)^\mu \quad . \tag{5.1}$$

This entropy has a simple meaning. Let us call $\Sigma(L)$ the number of steps required (on the average) to go through one confinement region (size L). This corresponds to

$$M_0(\Sigma) = L \quad ,$$

$$\Sigma = L^\mu \quad , \tag{5.2}$$

and the entropy reduction is simply

$$S(N, L) = \frac{N}{\Sigma(L)} \quad . \tag{5.3}$$

This S measures the quality of the reading: if $\Sigma(L = N)$ is comparable to N the sequence must be roughly maintained. But if $\Sigma(L = N)$ is smaller (large entropy) we loose all information.

For the free flights which are relevant below $d = 3$, we have

$$S(N, N) = N^{2 - d/2} \quad , \tag{5.4}$$

and S is large (e.g., $S \sim N^{1/2}$ for $d = 3$). Thus in the interval $2 < d < 3$ there is no sequential reading.

(b) *In the intermediate regime* $(3 < d < 4)$ we can again define a number of steps $\Sigma(L)$ — required to travel over a distance L — by the condition:

$$M(\Sigma) = L \ . \tag{5.5}$$

Here $M(L) \sim L^\nu$ and thus

$$\Sigma(L) = L^{1/\nu} \ . \tag{5.6}$$

We conjecture that the entropy reduction $S(N, L)$ at fixed L, variable N, is still *linear in N* as it is for other situations with excluded volume and confinement (de Gennes, 1979)

$$S(N,L) = \frac{N}{\Sigma(L)} \ . \tag{5.7}$$

The quality of sequential reading is still related to the quantity

$$S(N,N) = N^{1-1/\nu} \ . \tag{5.8}$$

Since $\nu < 1$ in this regime, $S(N,N)$ is still larger than unity, and we do not have sequential reading. But near $d = 4$ the situation improves: $S(N,N)$ is getting smaller.

(c) *For $d > 4$ $(\nu = 1)$* we have $M = N$, giving the correct sequence except for some noise. This noise corresponds to the coarse graining discussed in Sec. 2.

6. Conclusions

(a) For all $d > 2$ there is a coarse grained image which is reproduced by the reading chain.
(b) The printed sequence is read with the right sequential order (apart from weak noise effects) only for $d > 4$.
(c) The qualitative behavior seems to be the same for all finite values of the mismatch parameter x $(0 < x < \infty)$.
(d) Returning to more practical aspects: the present model is utterly unrealistic; most of the interesting action takes place at nonphysical dimensionalities $(3 < d < 4)$. But many extensions to more useful problems can be conceived: (i) incorporation of an excluded volume effect (which strongly reduces the weight of the large loops); (ii) unequal chemical distance S_i between active groups; (iii) faults in the chemical sequence of the reading chain; (iv) chemical

structures which are not linear chains; (v) partial rigidity of the connecting links. It may be that, by a suitable choice of templates we can arrive at systems which are efficient in reading and yet relatively simple to synthesize.

Acknowledgments

Most of this paper was written during visits at the departments of Chemical Engineering, Stanford University, the University of Delaware, and the Workshop on Finely Divided Matter (les Houches). Discussions with H. Nakanishi, G. Toulouse, P. Pincus, T. Witten, and J. Vannimenus have been of great help.

Appendix A: Weight of a Single, Long Loop

We consider here the diagram of Fig. 3, corresponding to a single long loop in the print which is read in the wrong order. The diagram has two long arcs of length σ, and we are interested in the asymptotic form of the weight $W(P)$ for large σ. The (not yet reduced) weight is

$$\langle W(P) \rangle = \int d\mathbf{a}_0 \, d\mathbf{a}_1 \ldots d\mathbf{a}_\sigma \exp -\tfrac{1}{2}(S_0 + S_1) \, , \qquad (A.1)$$

where

$$S_0 = \beta_0 (\mathbf{a}_0^2 + \ldots + \mathbf{a}_\sigma^2) \, ,$$

$$S_1 = \beta_1 \left[(\mathbf{r}_\sigma - \mathbf{r}_0)^2 + (\mathbf{r}_{\sigma+1} - \mathbf{r}_1)^2 + \mathbf{a}_1^2 + \ldots + \mathbf{a}_{\sigma-1}^2 \right] \, . \qquad (A.2)$$

The integrals in (A.1) break out into d factors, corresponding to the various component of the vectors. Thus we may put

$$\langle W(P) \rangle = I^d \, . \qquad (A.3)$$

Restricting our attention to one component $(\mathbf{a}_i \to a_i)$ we introduce a new variable u (corresponding to the vector length of the loop)

$$u = a_1 + \ldots + a_{\sigma-1} \, . \qquad (A.4)$$

Then the exponent becomes

$$S_1 = \beta_1 \left[(\nu + a_0)^2 + (\nu + a_\sigma)^2 + a_1^2 + \ldots + a_{\sigma-1}^2 \right] \, . \qquad (A.5)$$

After integration, and in the limit of large σ, we get

$$I = \left(\frac{\beta_1 - \beta_0}{2\beta_1} \right)^{1/2} \left(\frac{2\pi}{\beta_1 + \beta_0} \right)^{(\sigma+1)/2} \frac{1}{(\sigma-1)^{1/2}} \quad . \tag{A.6}$$

The corresponding factor for the exact sequence is

$$\left(\frac{2\pi}{\beta_1 + \beta_0} \right)^{(\sigma+1)/2} \quad .$$

Thus the reduced weight (as defined in Eq. 2.6) is

$$\widetilde{W}(P) = \left(\frac{1+x}{2\sigma} \right)^{d/2} \qquad (\sigma \gg 1) \quad . \tag{A.7}$$

This is not at all what we would have from the independent arc approximation

$$\widetilde{W}_{\text{ind}}(P) = h_x^2(\sigma) = \left(\frac{1+x}{\sigma} \right)^d \quad . \tag{A.8}$$

The source of the difference is clear: in the independent arc approximation $\mathbf{r}_\sigma - \mathbf{r}_0$ and $\mathbf{r}_{\sigma+1} - \mathbf{r}_1$ are decoupled; but for the loop of Fig. 2 they are strongly coupled. Whenever the loop is closed (and small) the two vectors $\mathbf{a}_0$ and $\mathbf{a}_\sigma$ take small, acceptable values. Thus we need only one factor $\sigma^{-1/2}$, and not two.

The discrepancy between (A.7) and (A.8) does not mean, however, that the independent arc approximation is bad: for the case of principal interest ($d < 4$) we have a large number of cross points in the tube, and the dominant structure is made with diagrams involving many interacting loops: then the notion of independent arcs is a plausible starting point. Above $d = 4$ the problem disappears altogether (there are no long loops).

Appendix B: Elastic Energy of an Ideal Levy Flight

An ideal flight of N steps has a certain scaling size $M_0(N)$ given by Eq. (4.8). We think of this flight as being a random chain, and we apply to the ends of the chain a certain extensional force f. The chain then takes a larger size M. We want to find the reduction S of the chain entropy, following a very physical argument by Pincus (1976). The idea is that the force f distorts the chain at large scales, but not at small scales. We can divide the chain into subunits of g monomers, such that the elongation X of one subunit (under the form f) gives an elastic energy of order kT (or an entropy loss of order unity). Thus, in reduced units

$$Xf = 1 .$$ (B.1)

At scales smaller than one subunit, the ideal flight is preserved

$$M_0(g) = X .$$ (B.2)

The overall elongation M of a chain of N monomers is the sum of N contributions from the subunits

$$M = \frac{N}{g}X$$ (B.3)

and the total entropy loss is

$$S = \frac{N}{g} .$$ (B.4)

We then insert Eq. (4.8) for $M_0(g)$ into (B.2), eliminate X between (B.2) and (B.3), and arrive at

$$S = \left(\frac{M}{M_0}\right)^{\mu/(\mu-1)} .$$ (B.5)

This whole picture holds if $(N/g > 1)$ (many subunits) or $M > M_0$.

References

Beardwood, J., Halton, J. and Hammersley, J. (1959) *Proc. Camb. Phil. Soc.* **55**, 299.

Birshtein, T., Ptitsyn, O. (1966) *Conformations of Macromolecules* (Wiley, New York).

de Gennes, P. G. (1979) *Scaling Concepts in Polymer Physics* (Cornell University Press, Ithaca).

Flory, P. (1971) *Principles of Polymer Chemistry* (Cornell University Press, Ithaca).

Grassberger, P. (1985) (to be published).

Halley, J. W. and Nakanishi, H. (1985) *Phys. Rev. Lett.* **55**, 551.

Hughes, B., Schlesinger, M. and Montroll, E. (1982) *J. Stat. Phys.* **28**, 111.

Kirkpatrick, S. and Toulouse, G. (1985) *J. Physique* **46**, 1277.

Pincus, P. (1976) *Macromolecules* **9**, 386.

Vannimenus, J., Mezard, M. (1984) *J. Physique Lett.* **45**, L-1145.

Westlaufer, D. B., ed., (1984) "The Protein Folding Problem," in *AAAS Selected Symposium 89* (Westview Press, Boulder).

PHYSICS OF MESOSCOPIC SYSTEMS

Yoseph Imry

School of Physics and Astronomy
Tel Aviv University
Tel Aviv, 69978
ISRAEL

and

IBM T. J. Watson Research Center
Yorktown Heights, NY 10598
USA

1. Introduction

Much of solid state theory and statistical physics is concerned with the properties of macroscopic systems. These are often calculated using the "thermodynamic limit" (system's volume Ω, and particle number N, tending to infinity with $n \sim N/\Omega$ constant) which is a convenient mathematical device for obtaining bulk properties. Usually, the system approaches the macroscopic limit once its size is much larger than some correlation length, ξ. In most cases ξ is on the order of a microscopic length (e.g., $\sim n^{-1/3}$), but in some special cases, such as in the vicinity of a second-order transition, ξ can become very large and one may observe behavior which is different from the macroscopic limit for a large range of sample sizes (Imry, 1969; Imry and Bergman, 1971; Fisher, 1971). Another case where the effective length scale dividing microscopic from macroscopic behavior is very large, is that of small conducting (or semiconducting) systems at low temperatures. Here, once an electron can propagate across the whole system without inelastic scattering, its wave function will maintain a definite phase and it will, thus, be able to exhibit a variety of novel interesting interference phenomena. In this paper we shall concentrate on the study of the latter type of systems.

The interest in studying these systems in the intermediate size range between microscopic and macroscopic — sometimes referred to as the "mesoscopic" (a word coined by Van Kampen, 1976) range — is not only in order to understand the macroscopic limit — and how it is achieved by, say, building up larger and larger clusters to go from a "molecule" to the "bulk." The special phenomena that exist in this range are of great interest in themselves. We shall see how fundamental principles of quantum mechanics (related to the concept of the phase of the wave function) and statistical physics (brought about by the lack of inelastic scattering and thermalization) appear and are amenable to theoretical clarification and experimental examination in these systems. Another interesting aspect is the distinction (Landauer, 1970; Azbel, 1973; Imry, 1977; Anderson *et al.*, 1980; Azbel and Soven, 1983) between ensemble-averaged properties and those specific to a particular given small system prepared under the same macroscopic constraints as with all the ensemble members. The specific "fingerprint" of such a small system is of interest and may be used to obtain some statistical information on the particular arrangement of the constituents in the system (Azbel, 1973). Many of the usual rules that one is used to in macroscopic physics may not hold in "mesoscopic" systems. For example, the rules for addition of resistances, both in series (Landauer, 1970; Anderson *et al.*, 1980) and in parallel (Gefen *et al.*, 1984a, b) are different and more complicated. The electron motion is wave-like and is not unsimilar to that of electromagnetic radiation in waveguide structures, except for complications due to disorder. These effects may set fundamental limits on how small various electronic devices can go. On the other hand, ideas for new devices, such as those operating in analogy (Fowler, 1985; Datta *et al.*, 1986b) with various optical and waveguide ones, as well as with SQUIDs (Superconducting Quantum Interference Devices), and other Josephson-effect systems (Hahlbohm and Lübbig, 1985), may emerge for small normal conductors.

The technology (see, e.g., Howard and Prober, 1982; Prober, 1983 and Laibowitz, 1983) for the fabrication of supersmall structures, using advanced optical or x-ray lithographic techniques, as well as electron-beams, is advancing very quickly, and has reached the stage where many theoretical predictions can now be confronted by experiments. Especially in semiconducting systems based on MOSFET or quantum well concepts, an excellent restriction in one direction exists (Ando *et al.*, 1982), so that creating small structures parallel to the 2D (two-dimensional) layer may achieve systems with a rather small number of active electrons or quantum states. One may soon reach the stage of having large conducting artificial molecules on which macroscopic experiments can be performed, in the same size range of ordinary macromolecules. The latter are of course of great interest too. It should be noted that photons may also be well "guided" in such systems and similar phenomena may thus occur for these electromagnetic waves, not to mention ideas for electron-photon coupling in various combinations.

In Sec. 2 we shall consider interference phenomena in the static properties of mesoscopic-scale conductors. The transport phenomena will be discussed in Sec. 3 using conventional methods. The Landauer-type approach for transport, which is particularly suited to these systems, will be developed and its applications treated in Sec. 4. Section 5 will give some concluding remarks, recent developments, notably the Hall effect (Büttiker, 1985a, b; Entin-Wohlman *et al.*, 1986) and universal conductance fluctuations (Altshuler, 1985; Lee and Stone, 1985; Altshuler and Khmel'nitskii, 1985; Imry, 1986) and discuss some (not necessarily electronic) fluctuation effects and open problems.

2. Static Electronic Properties of Mesoscopic Systems

2.1. *General*

When treating bulk systems, one is accustomed to use the simplification of a quasicontinuum of states. For a metallic "particle" with, say, 10^5 atoms the typical separation, w, between states at the Fermi energy is on the order of a few tenths of a degree K. Thus, clearly at temperatures of a degree K or less — this "graininess" of the levels becomes important and may influence, for example, the thermodynamic properties of the system, such as the specific heat or the magnetic susceptibility (Kubo, 1962; Gor'kov and Eliashberg, 1965; Mühlschlegel, 1983). The *precise* spectrum of such a system will usually depend on many details such as the shape or morphology of the surface of the grain, but for many applications, especially those concerned with an ensemble of grains, it is enough to have some statistical information on the level distribution. Powerful theories of these distributions in effectively random (due, e.g., to the high sensitivity to many uncorrelated details) systems exist and have been extremely successful in atomic and nuclear physics. These methods and their applications are thoroughly reviewed (Wigner, 1951, 1955; Dyson, 1962; Mehta and Dyson, 1963; Mehta, 1967), so we shall not discuss these aspects here in any detail, although one may feel that they should become very relevant for granular (Mühlschlegel, 1983) as well as strongly localized (Sivan and Imry, 1986b) condensed matter systems (a special application of these methods will be mentioned in Sec. 5). Neither shall we review the very interesting effects of resonances (Lifschitz and Kirpichnikov, 1979; Azbel and Soven, 1983) due to the discrete levels of the system nor those due to special hopping paths (Lee, 1984). The effects we shall mainly concentrate on in this paper have to do with the interference of the electron waves and, in particular, its sensitivity to magnetic fields or fluxes (Aharonov and Bohm, 1959). Such interference phenomena exist in principle when the temperature is low enough so as not to disturb the coherence of the wave functions over the relevant spatial scales.

Our discussion of these effects will consist of two parts. First, equilibrium properties, such as the energy, magnetization, and magnetic susceptibility will be discussed in this section. Then we shall consider transport properties in the next two sections. In both cases the scattering of electrons by impurities, defects, imperfect surfaces, etc., will play an important role but will not eliminate the interference.

Some of the interference phenomena that we shall discuss have some analogy to effects (e.g., related to flux quantization) that are well known and documented in superconductors, where they are brought about by the appropriate "off diagonal" long range order which exists there (Byers and Yang, 1961; Bloch, 1970). However, we emphasize that we shall consider here only *normal conductors*, where the possibility of coherence is not related to electron correlation but simply to the (finite) size of the sample being smaller than the appropriate phase randomization length.

A further consideration which may limit the *magnitude* of the phenomena under discussion is the possibility that while they exist, they may experience some averaging due, for example, to the electrons not being monoenergetic because of the finite temperature. It will turn out that smaller sizes and/or lower temperatures increase the magnitude of the effects, to the point where they are not only relatively easily observable but also potentially applicable, as mentioned in the introduction.

2.2. Quantum interference effects in equilibrium properties

It has first been noticed by Dingle (Dingle, 1952) that equilibrium properties such as the average energy or magnetization of a small *free electron* system with a simple, *ideal,* geometry, e.g., a perfect disc or ring, are sensitive to a magnetic field. Oscillatory behavior is obtained as a function of the field, where the scale is set by the magnetic flux through the system being on the order of a flux quantum, $\Phi_0 = hc/e$. In the particular case of a ring (see Fig. 1a) with an Aharonov-Bohm (Aharonov and Bohm, 1959) flux Φ through its opening, the thermodynamic functions are periodic in Φ with a period Φ_0.

We shall from now on first concentrate in this section on the ring geometry. Dingle's type of results have been obtained later by several researchers in different contexts (Gunther and Imry, 1969; Kulik, 1970; Brandt *et al.*, 1976, 1982). However, an important difficulty looming in the background has been that electron scattering was almost universally expected (Kulik, 1970; Altshuler *et al.*, 1981) to eliminate these effects in any realistic system.

The point is that it is very difficult to expect a real system to be not only impurity- and defect-free but to also have perfect surfaces. Some surface roughening (perhaps not as bad as in Fig. 1b) will practically always exist. Thus, the elastic mean free path, l, will be limited at best by the ring arms' width and thickness. Hence l will be typically much smaller than, say, the ring's circumference L — which is the distance over which the electron's wavefunctions experience interference. One's first intuition would say that the many scatterings the electron has to experience when travelling around the ring would completely eliminate any interference pattern. This is, in fact, in agreement with common notions on electron beam diffraction experiments (including Aharonov-Bohm type, Chambers, 1960; Merzbacher, 1961; Tonomura *et al.*, 1982) where care has to be taken to perform the experiment in a high enough vacuum to reduce the electron random scattering.

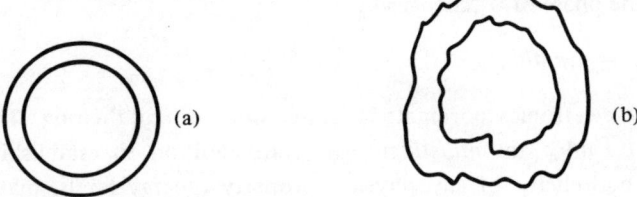

(a)   (b)

Fig. 1. Schematic ring geometries: (a) ideal ring; (b) ring with severely rough shape and surface.

We shall see, however, that the above expectation is, in principle, wrong and the analogy with beam experiments misleading. The point is that there exists an important distinction between *elastic* scattering, due to some *static* potential — in which wave functions with well-defined phases exist, and *inelastic* scattering. In the latter case, the electron may excite a phonon or alter the state of a "dust" particle, etc. The electron will as a result not have definite energy and phase, and some practical irreversibility is introduced (since the time-reversed process of sending the phonon back exactly in the correct way, while it is possible in principle, is not practically relevant if the phonon in turn interacts with a large assembly of, say, thermal-bath phonons). The important distinction between the effects of elastic and inelastic scattering has become clear through the recent understanding (Thouless, 1977; Bergamann, 1984; Lee and Ramakrishnan, 1985; Imry, 1983) of conduction in disordered systems via localization theory (Anderson, 1981; Abrahmas *et al.*, 1979). Prior to that, R. Landauer (1966) has informally expressed similar insights (based on Landauer, 1957) and Gunther and Imry (1969) found persistent diamagnetic currents in a system with a finite resistance. Before discussing this in more detail we briefly review some general background results for "rings."

Consider a general doubly-connected system with an Aharonov-Bohm flux Φ through its opening. An important and very general theorem due to Byers and Yang (1961) and Bloch (1970) states that all physical properties of this "ring" are periodic in Φ with a period Φ_0. The proof proceeds by eliminating Φ with the gauge transformation

$$\psi' = \exp\left[\frac{ie}{\hbar c} \sum_j \chi(r_j)\right] \psi \quad , \tag{2.1}$$

where r_j are the electron coordinates and χ is defined by $\mathbf{A}_\Phi = \nabla\chi$, where $\mathbf{A}_\Phi$ is the vector potential whose curl is the Aharonov-Bohm magnetic field (i.e. curl $\mathbf{A} = 0$ in the material and $\oint \mathbf{A}_\Phi \cdot d\mathbf{l}$ on a path circulating the ring's opening is equal to Φ). The gauge-transformed many-electron Schrödinger equation has $\mathbf{A}_\Phi = 0$. The price for this is, of course, that the transformed wave function, ψ', does not in general satisfy periodic boundary conditions around the ring. In fact, the phase of ψ' changes by

$$\Delta\phi = 2\pi\Phi/\Phi_0 \quad , \tag{2.2}$$

when one electronic coordinate is rotated once around the ring. Thus the fluxes Φ and $\Phi + n\Phi_0$ are *indistinguishable*. In addition to establishing the exact claimed periodicity of any physical property (energy levels, matrix elements, etc...), Eq. (2.2) also tells us that a noninteger flux is mathematically equivalent to a change in boundary conditions. This concept will prove extremely useful to us and we shall discuss an even more elementary way to establish it later. In any system obeying the *classical* laws, the Aharonov-Bohm flux Φ is clearly irrelevant. All physical properties do *not* depend on it. This is, of course, trivially consistent with the above theorem (a constant is also periodic, but not very interesting). The real issue is whether there is a sizeable sensitivity of, say, the energy levels or the transition probabilities of Φ. (Periodicity is always guaranteed!) Here, one can make connection (Büttiker *et al.*, 1983) with an important idea due to Thouless (1977) and Edwards and Thouless (1972). According to this the conductance of a system is on the order of $e^2/\hbar$ times the (dimensionless) ratio between the sensitivity V, of its energy levels to changes in the boundary condition, to the separation w, between levels (both at the Fermi energy E_F). Thus, knowledge of the conductance of the system will enable us to estimate its flux sensitivity (Kohn, 1964). We shall review and have some remarks on the Thouless criterion in Sec. 3.

Büttiker *et al.* were the first (Buttiker, Imry and Landauer, 1983) to consider this issue using the simple model of a one-dimensional (1D) ring with disorder. They noted that the boundary condition in Eq. (2.2) is similar to that satisfied by the Bloch function ψ_k in a periodic potential across the unit cell of size L. Thus, identifying $\Delta\phi$ with kL establishes a one-to-one correspondence between the two problems. In fact, the condition in Eq. (2.2) can then be understood since the electron experiences the same potential by moving again and again around the ring, i.e., an effectively periodic situation, where the whole circumference of the ring plays the role of the unit cell. The electronic energy levels of the ring as functions of Φ are schematically depicted in Fig. 2a. Note that this schematic form is applicable for an arbitrary random potential along the ring, since it can be shown (see Peierls, 1955) that in 1D the only extrema of $E(k)$ are at $k = 0$, $\pm\pi/a$. For nearly free electrons one gets the usual wide bands and narrow gaps (i.e. $V \sim w$) while for a strong potential (small transmission along the ring) the opposite tight-binding situation (narrow bands, large gaps) i.e., $V \ll w$ is obtained. The latter case corresponds to strong localization at E_F. It is possible to estimate the flux dependence of the total energy E at low temperatures ($k_B T \lesssim w$), since it is the sum of all occupied levels. Due to the alternating sings of $\partial E/\partial\Phi$ for consecutive levels, one has a strong cancellation and the sum is on the order of the last term, around E_F. The circulating current is given (Byers and Yang, 1961; Bloch, 1970) by $c^{-1}(\partial E/\partial\Phi)$. Thus, for N electrons, assuming $V \sim w$

$$I = \frac{1}{c}\frac{dE}{d\Phi} \sim \frac{eE_F}{Nh} ,\qquad\qquad (2.3)$$

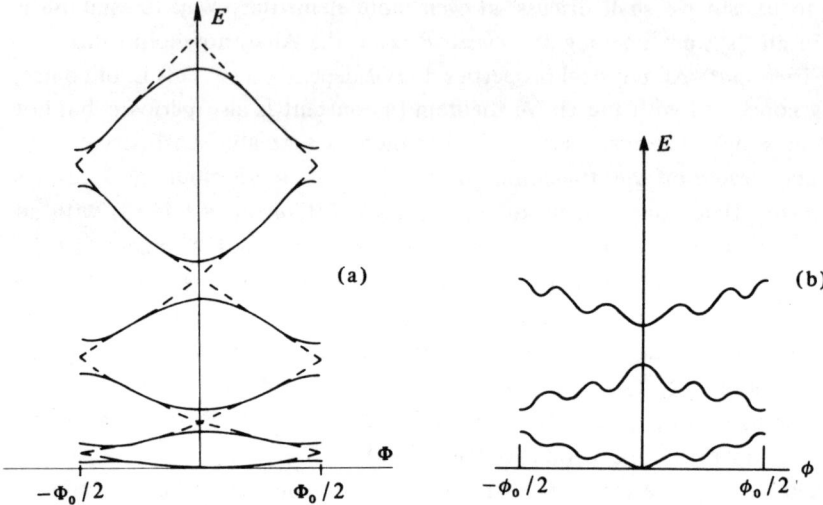

Fig. 2. Energy levels as functions of flux, Φ, for a (a) 1D ring (schematic); (b) non-1D ring.

where E_F is measured in electron-volts. For a strictly 1D ring made from metal with $E_F \sim 2$ eV with a circumference of a micron, $I \sim 10^{-8}$ amp. Since $w \sim 10$K having $k_B T < w$ is quite feasible. A further condition necessary to observe the oscillation is a long enough inelastic time, i.e.,

$$\hbar/\tau_{in} \ll w, V \quad , \tag{2.4}$$

namely, that the level width is much smaller than the level separation or the bandwidth, whichever is smaller. The condition $\hbar/\tau_{in} \ll V$ can be shown via the Thouless criterion to be equivalent to $\sqrt{D\tau_{in}} \gg L$. An unfavorable case is the limit $V \ll w$, where the effects are very small — localized states are not sensitive to boundary conditions.

Note that in the presence of a finite nonintegral Φ, the small diamagnetic-type currents flowing around the ring are *persistent* once the temperature is low enough so that Eq. (2.4) is satisfied. The currents do *not* decay if the inelastic scattering is weak enough — the latter just establishes, within a few τ_{in}, an equilibrium population among the states and the current is given by the appro-priate average, but there is no way for the persistent currents to decay. This result is quite surprising but correct. It is actually of the same nature as the persistence of the usual diamagnetic currents in bulk samples. The decrease of this equilibrium current amplitude with decreasing τ_{in} and increasing dissipation has been discussed by Landauer and Büttiker (1985) and Büttiker (1985c). These currents yield an orbital magnetic moment M (sometimes referred to as "diamagnetic," although M may be parallel or antiparallel to H) and magnetic susceptibility χ oscillating as functions of Φ, with a period Φ_0.

So far we discussed the case where all the magnetic fields were pure Aharonov-Bohm type. In the case where there are also some nonzero fields inside the metal itself leading to a flux Φ_M, their (paramagnetic as well as orbital) effects have to be added. There will be no periodicity in the total magnetic flux, only in the dependence on the Aharonov-Bohm part Φ, but that should be on top of the effects due to the magnetic fields inside the material. If the ratio of the area of the hole to that of the material is large enough ("good aspect ratio") one may expect the fast periodic dependence on Φ to be still visible on top of the slower variation due to Φ_M.

Many interesting questions occur when the flux Φ changes with time to yield an e.m.f. $V = -(1/c)(d\Phi/dt)$. For the case where V is pure dc, the resulting current will oscillate (Bloch, 1970) with a Josephson frequency

$$\omega = eV/\hbar .\tag{2.5}$$

When the change of Φ is not slow enough (for an ac voltage or for a finite dc one), Zener type transitions may occur among the bands, which necessitates a dynamical treatment which we shall not touch upon here.

Up to now, we discussed only the pure 1D case. However, in most conceivable experiments the wires making the ring have a finite cross section A. Thus, the number of transverse states (across the wire) below E_F is on the order of

$$N_\perp \sim k_F^2 A ,\tag{2.6}$$

and the total number of electrons is

$$N \sim k_F L N_\perp .\tag{2.7}$$

Here, the levels as functions of Φ display a much more complicated structure with many maxima and minima, than in the 1D case. Schematically, the structure is more like Fig. 2b rather than Fig. 2a. It is not straightforward to estimate even the order of magnitude of the "persistent" diamagnetic current. The most pessimistic estimate will yield something like Eq. (2.3), with N replaced by Eq. (2.7) and with possible further reduction for $T \gg w$ (which will almost always happen, since $w \sim E_F/N$). This would make the observation of the oscillating "diamagnetic" currents rather difficult although still possible. However, it is not obvious that the situation is really so bad. In fact, another estimate, based on the assumption that $dE_i/d\Phi$ are randomly positive or negative, *with no correlation*, will yield a $1/\sqrt{N}$ dependence of the total current on N. To make this estimate slightly more quantitative, we note that, via the

Thouless argument, and since $|dE_i/d(\Delta\phi)|$ is on the order of the Thouless V:

$$\left|\frac{dE_i}{d\Phi}\right| \sim \frac{\hbar}{\Phi_0 Re^2} \frac{E_F}{N} \, , \tag{2.8}$$

with R being the resistance of the wire making, say, one-half of the ring. Since N such contributions add with random signs, we find, at low temperatures

$$I = \frac{d\langle E\rangle}{d\Phi} \sim \frac{E_F}{eR} \frac{1}{\sqrt{N}} \, . \tag{2.9}$$

For a metallic wire of $1\mu \times (500\,\text{A})^2$ and $R = 100\,\Omega$, we find $I \sim 5\mu\text{A}$. Thus, *if* this estimate is valid, the persistent currents and the associated susceptibility would be quite observable. Note that even in the $O(1/N)$ estimate, one should use Eq. (2.8) which will then give the current amplitude. Obviously, better estimates for these currents are necessary.

It should, however, perhaps be emphasized that such persistent currents do, in fact, occur on the surface of any ordinary bulk metallic diamagnet in a finite magnetic field. According to the usual picture, the local microscopic diamagnetic currents cancel inside an homogeneous sample, but a nonzero surface contribution remains (an interesting case is that of a nonuniform system — e.g., a mixture of a metal and an insulator — where nonzero diamagnetic currents may exist in the bulk — e.g., along the metal-insulator interfaces). In the ring geometry, these currents flow along the inner and outer surfaces, and the novel property is the periodicity of the total circulating current as a function of Φ.

3. Quantum Interference Effects in Transport Properties

3.1. *Generalities on Sections 3 and 4*

The transport properties of mesoscopic systems display a wealth of interesting phenomena that are quite novel with respect to the usual macroscopic systems. In this paper we shall first consider in some detail the periodic oscillations in the magnetoresistance of a ring as a function of the Aharonov-Bohm flux, and then briefly discuss the aperiodic but analogous fluctuations (Umbach *et al.*, 1984; Jackel, 1983; Blonder, 1984; Skocpol *et al.*, 1986; Licini, Dolan and Bishop, 1985; Licini, Bishop, Kastner and Melngailis, 1985; Stone, 1985; Altshufer, 1985; Lee and Stone, 1986) in a fine singly connected wire. As in the previous section, there is an important distinction here between the effects of elastic and inelastic scattering. Also, since the system is so small, its measured resistance may depend on the existence, type and structure of contacts made onto it. The usual rules for connecting resistances in series and in parallel may break down in the "quantum" regime. Various other effects, familiar to varying extents from waveguiding systems may also occur. For example: An open-ended branch can greatly change the resistance of the system (Gefen *et al.*, 1984a, b); the resistance may be nonlocal in the sense that what is measured between a given pair of points may depend on things connected further away (Anderson *et al.*, 1980; Engquist and Anderson, 1981); there exist contact and spreading resistances which are not always negligible.

From the theoretical point of view, this problem is also interesting due to the electrons being possibly further removed from equilibrium (due to the scarcity of inelastic scattering) than in ordinary transport theory. In some cases, one has to develop special methods to handle such aspects. In this section, we shall review the Kubo linear response formalism (Kubo, 1957, 1962) and its subtleties for our case, and apply those insights to the Thouless picture of conductance in disordered systems. In the next section, we shall develop the Landauer-type formulation for the conductance of a segment of a disordered system between two ideal leads. The similarities and differences between these two approaches will be discussed as well.

3.2. The Kubo formulation for mesoscopic systems

For an infinite system, the conductivity at frequency ω may be obtained by calculating, using the golden rule, the power absorbed from a classical e.m. field by the system (we shall always consider here the σ_{xx} component). The field used in this formulation is the actual one, often different from the applied one (see e.g. Landauer, 1977)

$$ \sigma(\omega) = \frac{1}{\Omega} \frac{2\pi}{\omega} e^2 \sum_{k,l} |\langle k | \hat{v}_x | l \rangle|^2 \, \delta(E_l - E_k - \hbar\omega)(f_k - f_l) \; ; \qquad (3.1) $$

for simplicity, we consider noninteracting (or Hartree-Fock) electrons. Ω is the volume of the system; $|k\rangle$, $|l\rangle$ are the free (or self-consistent single) electron states and f_k, f_l their populations, $\hat{v}_x$ is the velocity operator of the electron in the x direction. The assumption of an infinite system is crucial here, in order to have a continuum of states. Otherwise, the field does not induce real transitions. An *isolated* finite system with a truly discrete spectrum does *not* in fact really absorb energy from the field. In order to obtain a finite conductivity, the small system has to be (and in real situations is) coupled to a very large heat bath, for example, an assembly of thermal phonons. This enables energy to be transferred from the e.m. field into the bath via the small electronic system. For a weak enough interaction with the bath, one may say that the discrete levels of the system acquire finite widths. It then makes sense to write down Eq. (3.1) with E_k having a finite width or with an imaginary part $i\eta$ to the frequency ω. Thus, for d.c. (Re $\omega \to 0$) Thouless and Kirkpatrick (1981) following Czycholl and Kramer (1979), suggest the following expression for σ_{dc}:

$$\sigma(i\eta) = \frac{1}{\pi} \int_{-\infty}^{\infty} \frac{\sigma(\omega')\eta}{\omega'^2 + \eta^2} d\omega'$$

$$= \frac{2e^2\hbar}{\Omega} \sum_{k,l} \frac{|v_{kl}|^2}{E_k - E_l} \cdot \frac{\hbar\eta(f_k - f_l)}{(E_k - E_l)^2 + (\hbar\eta)^2} \quad . \tag{3.2}$$

While this procedure certainly makes sense in smoothing out the δ functions of Eq. (3.1), it does need a more rigorous justification in terms of the combined electron-bath system. Van Vleck and Weisskopf (1945) obtained similar results using a semiclassical picture with collision broadening, which is discussed further in Imry and Shiren (1986), and the following discussion will be based on Eq. (3.2).

It is possible to show that once $\hbar\eta$ is much larger than the energy level separation, w, of the electrons at E_F, Eq. (3.2) goes over, as it should, to the appropriate bulk expression. This condition is always very well satisfied for macroscopic systems where $w/k_B \sim 10^{-18}$ K and $\hbar/\tau_{in}k_B$ is rarely smaller than $\sim 10^{-4} - 10^{-5}$ K (and it usually attains such values only at mK temperatures). To get the usual expression from Eq. (3.1), one straightforwardly obtains for the low temperature conductivity, by replacing the sums by integrals and assuming that $|\langle l|\hat{v}|k\rangle|^2$ has some typical value denoted by $|\langle v\rangle|^2$ near E_F,

$$\sigma_{KG} = 2\pi e^2 \Omega \hbar |\langle v\rangle|^2 [n(0)]^2 \quad , \tag{3.3}$$

where $n(0)$ is the density of states at E_F. We shall refer to this as the Kubo-Greenwood (Kubo, 1957; Greenwood, 1958) conductivity.

However, for the typical small metallic systems of interest to us here, w can become of the order of a few mK. Thus, at temperatures below ~ 0.1 K, one may get into an interesting novel range where

$$\hbar\eta \lesssim w \quad . \tag{3.4}$$

In the limit $\hbar\eta \ll w$ the conductivity is easily estimated (using $|E_k - E_l| \sim w \sim [n(0)\Omega]^{-1}$) to be on the order of

$$\sigma \sim \sigma_{KG} \frac{\hbar\eta}{w} \quad . \tag{3.5}$$

This has the interesting feature that the conductivity which is defined by energy absorption from the EM field vanishes (Landauer and Büttiker, 1985; Büttiker, 1985b; Imry and Shiren, 1986) when $\hbar\eta/w \to 0$. In this limit, we have discrete states with no real energy absorption. Thus, the d.c. conductivity looks as a function of η schematically like Fig. 3. We emphasize that this discussion was concerned with a particular definition of the conductivity, as measured for example by putting the sample, with no contacts, in an electromagnetic cavity and measuring the extra absorption due to the sample at low frequencies. It will turn out that this definition is *not necessarily identical* to others. For example, we shall find in Sec. 4 that the same sample will display a well-defined, finite, resistance which will be independent of η for small enough η, if measured by connecting to it two appropriate contacts. In that case the Joule energy dissipation will take place inside the thermal baths that have to be assumed to be associated with the contacts, as will be discussed there.

The Kubo-Greenwood formulation, which can be conveniently cast in terms of time-dependent correlation functions, has been an extremely useful way to do transport theory. It is also the basis for a systematic diagrammatic expansion in the strength of the disorder, characterized by the small parameter $(k_F l)^{-1}$, l being the elastic mean free path. The first correction (Langer and Neal, 1966; Gor'kov *et al.*, 1979; Abrahams *et al.*, 1979, Hikami *et al.*, 1980) to classical, Boltzmann, transport yields the weak localization contributions. These quantum corrections, due to coherent backscattering of the electron (Brandt *et al.*, 1976) (at $H = 0$) are amply presented and discussed in several excellent reviews (Altshuler *et al.*, 1982; Bergmann, 1984; Lee and Ramakrishnan, 1985; Fukuyama, 1983), and we shall not repeat these discussions here. Suffice it to say that this weak localization theory can explain and predict, sometimes even quantitatively, an extremely large amount of $\sigma(H, T)$ data on many totally

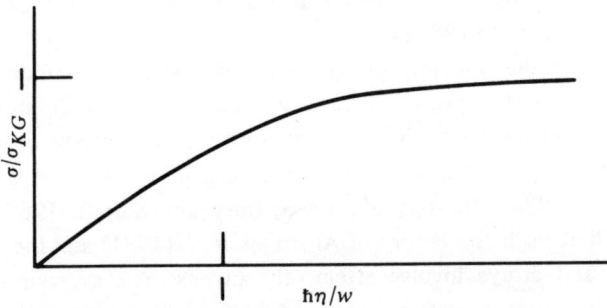

Fig. 3. The d.c. conductivity as function of the ratio $\hbar\eta/w$. It approaches the Kubo-Greenwood conductivity σ_{KG} for $\hbar\eta \gg w$, and is proportional to $\sigma_{KG}\hbar\eta/w$ for $\hbar\eta \ll w$.

different not-too-disordered systems. In the case of, for example, 2D thin films, MOSFET and quantum well systems, the corrections are quite universal and depend on a small number of parameters such as the sheet resistance, inelastic and spin-orbit scattering times.

One of the most interesting predictions of weak localization theory has been the one by Altshuler, Aronov and Spivak (AAS) (1981) on periodic oscillations of the (Kubo-type) conductance of small doubly connected samples (such as rings or small-radius cylinders), as a function of the Aharonov-Bohm flux Φ, through their opening. One surprising aspect of this calculation has been that the fundamental period of the oscillations was not Φ_0, as the general theorem alluded to in Sec. 2 predicts, but $\Phi_0/2$. The $\Phi_0/2$ period is the "second harmonic" of the Φ_0 oscillation, thus, this periodicity does not contradict the above theorem. The question is only why does the fundamental Φ_0 period not appear.

Before answering this question we mention that the prediction of the $\Phi_0/2$ oscillation following the beautiful pioneering work by Sharvin and Sharvin (1982) has received very convincing experimental support (Altshuler *et al.*, 1982b; Gijs *et al.*, 1984). In the more recent experiments on long cylinders an almost quantitative agreement with the full theory (taking into account the non-Aharonov-Bohm magnetic field inside the material) was achieved. The $\Phi_0/2$ oscillation has also been clearly seen in experiments on large arrays (Pannetier *et al.*, 1984. 1985; Bishop *et al.*, 1985; Licini, Dolan and Bishop, 1985; Licini, Bishop, Kastner and Melngailis, 1985; Dolan *et al.*, 1986) of many "rings". In all these experiments the fundamental, Φ_0 period has not been seen. Preliminary experiments (Umbach *et al.*, 1984; Webb *et al.*, 1984) on single rings were inconclusive, but did show traces of perhaps both Φ_0 and $\Phi_0/2$ oscillations, with additional aperiodic structure that will be discussed in Sec. 4. Only during the writing of this paper have convincing Φ_0-periodic oscillations in single small rings been reported (Webb *et al.*, 1985a, b; Washburn *et al.*, 1985; Chandrasekhar *et al.*, 1985; Datta *et al.*, 1986a).

The answer to the dilemma of where the Φ_0-periodic oscillation in many experiments and in the weak localization theories are, while motivated and amplified by work reviewed in the next section, can already been presented here. The point is (Gefen, 1984; Carini *et al.*, 1984; Browne *et al.*, 1984; Büttiker *et al.*, 1985; Murat *et al.*, 1986; Imry and Shiren, 1986; Stone and Imry, 1986) that both the theory of Altshuler *et al.* (1981) and the experiments on cylinders and arrays involve effectively an ensemble averaging over many microscopically distinct systems prepared with the same overall macroscopic conditions. Thus, all rings in the array have similar impurity concentrations but the precise configuration of the impurities is obviously different. In the perturbative theoretical calculations one ensemble-averages from the very beginning in order to use propagators that depend only on relative distances (apart from boundary effects). In the cylinder experiments, the resistance is measured *along* an approximately 1 cm long cylinder which consists of around 10^4 pieces of length L_Φ added classically in series. Now, the work on rings with contacts (next section) suggests that the Fourier coefficient corresponding to the Φ_0-periodic part of the oscillation does not have a definite phase. On the other hand the AAS $\Phi_0/2$ Fourier coefficient does have a definite phase (e.g., $G(\Phi)$ is minimal (maximal for systems with strong spin-orbit scattering) at the origin (Altshuler *et al.*, 1982b; Bergmann, 1984; Lee and Ramakrishnan, 1985), $\Phi = 0$). Thus, the ensemble averaging (*if done on a broad enough ensemble!*) eliminates the Φ_0 fundamental component, but the AAS $\Phi_0/2$ one survives. This idea is, of course, in agreement with the fact that later IBM (Webb *et al.*, 1984, 1985a; Washburn *et al.*, 1985; Webb *et al.*, 1985b) and Yale (Chandrasekhar *et al.*, 1985) experiments on *single* rings, triggered to some extent by the theoretical developments, have in fact displayed *large* distinct Φ_0 periodicities. More recently, this periodicity was also seen in small semiconductor heterostructures by Datta *et al.*, (1986).

These experiments had in fact been preceded by model calculations by Imry and Shiren (1986) on the Kubo conductivity of closed 1D rings and by Murat, Gefen and Imry (1986) on 1D rings with contacts. The latter will be mentioned in the next section. Results from the former are displayed in Fig. 4b, in which instead of ensemble averaging, the conductivity is calculated at a series of increasing temperatures T. It turns out that in 1D, once $k_B T \gg w$, different electrons in the "thermal band" of width $k_B T$ around E_F have different phases associated with propagation around the ring. Thus, high enough temperatures, where the relevant energy scale in 1D turns out to be the level separation w, provide self averaging which is similar to ensemble averaging. This will be discussed more in subsequent sections. However, a glance at Fig. 4 is enough to demonstrate the better and better averaging out of the Φ_0-periodic component with increasing T. The idea that ensemble averaging may lead to a $\Phi_0/2$ fundamental periodicity was also given and proven by Carini *et al.* (1984). However, since they considered a certain quantity (participation ratio) averaged over a *whole band*, they also found that the Φ_0-periodic component decreased like $1/L$ (L being the length of the system) even for a *single* ring. Averaging over the whole band is like thermal averaging with $k_B T$ comparable with the *electron band width*. This is not applicable to the low temperature conductivity.

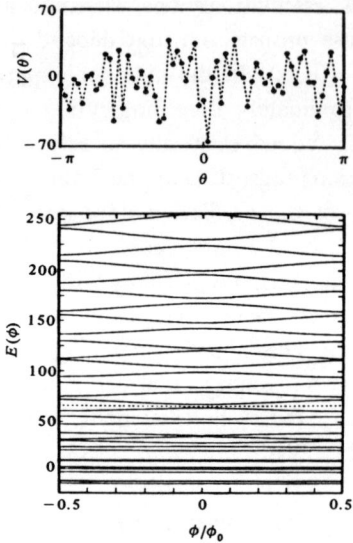

Fig. 4a. A random potential $V(\theta)$ (θ is the azimuthal angle) on a ring with its energy levels as functions of ϕ/ϕ_0. The dotted line is the Fermi energy.

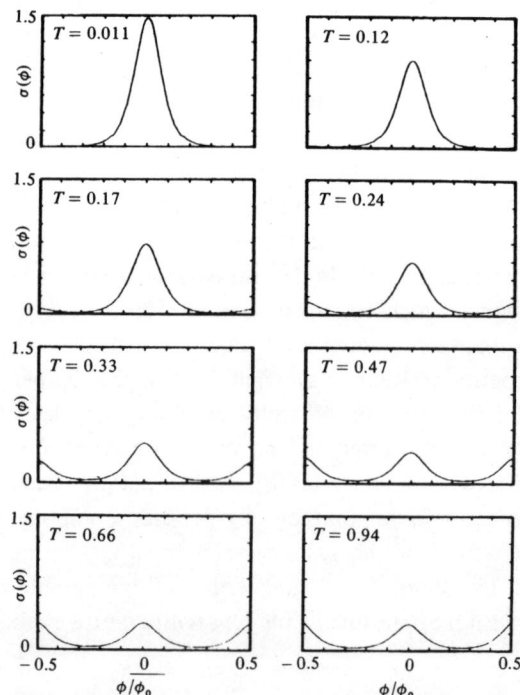

Fig. 4b. σ vs. Φ/Φ_0 for the above ring for $\eta = 0.037$ at indicated temperatures, exhibiting the crossover from Φ_0 to $\Phi_0/2$ periodicity with increasing energy averaging (from Imry and Shiren (1986)).

3.3. Discussion of the Thouless considerations and modifications in mesoscopic systems

Much of the physical basis of the modern scaling theory of localization (Abrahams *et al.*, 1979) has to do with ideas due to Thouless and his co-workers (Thouless, 1977; Edwards and Thouless, 1972). According to this, one considers, say, a "hypercube" in d-dimensions, of side L made of the given disordered system, or divides mentally the large system into many such hypercubes. If the diffusion time of an electron across this length L is τ_D, the diffusion constant is given by $D = L^2/\tau_D$. A typical energy, V, is now defined and related to τ_D by an uncertainty-type relationship, $V \sim \hbar/\tau_D$. One now calculates the conductivity σ from D, via the Einstein relationship

$$\sigma = e^2 D \frac{\partial u}{\partial \mu} \to e^2 Dn(0) , \qquad (3.6)$$

where the last equality is valid for noninteracting degenerate fermions, which is the case we consider here. The conductance, $G \sim \sigma L^{d-2}$ is now given by

$$G \sim \frac{e^2}{\hbar} \frac{V}{w} , \qquad (3.7)$$

where $w \sim (n(0)\Omega)^{-1}$ is the characteristic separation of levels at the Fermi energy. The energy V has to do with how the information from, say, the surface of the cube is propagated into its inside. Thus, V should be of the order of the sensitivity of the energy levels to boundary conditions. We recall, via the arguments of Sec. 2.2, that a change in boundary conditions, say, from periodic to antiperiodic is equivalent to putting the system on a ring and changing Φ/Φ_0 from integer to half-integer.

A more detailed justification of the important relationship, Eq. (3.7), was given by Edwards and Thouless (1972) via the Kubo-Greenwood expression (Eq. (3.3) above), to which Eq. (3.7) is equivalent. Thus, the condition

$$\hbar\eta = \hbar/\tau_{in} \gg w , \qquad (3.8)$$

also serves as a condition for the validity of the Thouless relation, Eq. (3.7).

We first mention some applications of this to temperature-dependent conduction in bulk systems, which is also an opportunity to very briefly review (Altshuler *et al.*, 1982a; Lee and Ramakrishnan, 1985; Imry, 1983) the scaling theory of localization (Abrahams *et al.*, 1979) and its application to conduction in disordered systems (Imry, 1980a, 1983a). This theory of localization (Abraham *et al.*, 1979) considers the change of the dimensionless conductance based on the RHS of Eq. (3.7),

$$g(L) = \frac{V(L)}{w(L)} \quad , \tag{3.9}$$

as function of L. The condition that g is also the dimensionless conductance $G\hbar/e^2$ (i.e., that Eq. (3.7) is valid) is, again Eq. (3.8). We now summarize the scaling predictions on the behavior of $g(L)$, neglecting electron-electron, spin-orbit interactions, etc.

For dimensions $d \leqslant 2$, the scaling theory predicts that the system is always insulating as $L \rightarrow \infty$. G is a monotonically decreasing function of L. If the system is heavily disordered (a good measure for that is the value of g, g_0, on some short-scale $L_0 \sim \max(l, k_F^{-1})$) in the sense that if $g_0 \ll 1$, then $g(L)$ always decreases exponentially with increasing L. For light disorder, $g_0 \gg 1$, the localization length, ξ, is defined by

$$g(\xi) = C \ , \tag{3.10}$$

where C is a numerical constant of order unity, so that for $L \gg \xi$, g is again exponential in $(-L)$. For $d > 2$, the situation is different. Now, there exists a localization transition at a critical value of g_0, g_c. For $g_0 < g_c$, the situation is similar to that at lower d, in the sense that the system is insulating, $g(L) \propto \exp(-L/\xi)$, for $L \gg \xi$, except that here $\xi \rightarrow \infty$ as $g_0 \rightarrow g_c$. For $g_0 > g_c$ the system is metallic and there is again a correlation length ξ (Imry, 1980a) above which $g(L) \sim \sigma_M \cdot L^{d-2}$, where σ_M is the macroscopic metallic conductivity (multiplied by $\sim \hbar/e^2$). Here, again $\xi \rightarrow \infty$ as $g_0 \rightarrow g_c^+$.

For bulk systems ($L \rightarrow \infty$) one proceeds by identifying a phase coherence length (Thouless, 1977), L_Φ, for the electrons, which is identical in most cases with the length, L_{Th} :

$$L_{Th}^2 = D\tau_{in} = D\hbar/\eta \ , \tag{3.11}$$

introduced by Thouless. Except in the localized phase with $L_{Th} \gg \xi$, where quantum diffusion proceeds only up to scale ξ, L_{Th} is the characteristic length over which the electron diffuses coherently (by pure quantum mechanics in the random potential) between inelastic collisions. If the latter are assumed to effectively randomize the phase of the electron's wave function, then $L_\Phi \cong L_{Th}$. Except in the localized phase with $L_{Th} \gg \xi$, one can use the pure localization theory on scales $L \lesssim L_{Th}$. On the other hand, on scales $L \gtrsim L_{Th}$, the classical laws should be valid. This suggests the following approximate procedure to evaluate the bulk conductivity. Localization theory is used up to scales $\sim L_{Th}$ to obtain $G(L_{Th})$. Then, classical resistance addition is used to go from $L \sim L_{Th}$ to the macroscopic sample size, i.e.,

$$\sigma_{\text{macro}} \sim G(L_{Th})/L_{Th}^{d-2} \ . \tag{3.12}$$

For $L_{Th} \lesssim \xi$, *both* in the localized and extended phase, $g(L_{Th})$ is a (disordered-dependent) number times $e^2/\hbar$. For $L_{Th} \gg \xi$ in the metallic phase, $G(L_{Th}) = G(\xi)(L_{Th}/\xi)^{d-2}$, so that $\sigma_{\text{macro}} \sim G(\xi)/\xi^{d-2}$, as is well known (Abrahams *et al.*, 1979). We should now examine the previously discussed condition for the validity of this procedure, namely that Eq. (3.8) is satisfied on the scale L_{Th}.

We find from $L^2 = D\hbar/V$, Eqs. (3.8) and (3.11) that the condition in Eq. (3.8) reads for scale L

$$L_{Th}^2/L^2 \lesssim V/w \ , \tag{3.13}$$

where the RHS is evaluated on scale L. For the scale L_{Th}, we find $V/w \gtrsim 1$, or, in terms of the dimensionless conductance g

$$g(L_{Th}) \gtrsim 1 \ . \tag{3.14}$$

This is indeed roughly valid for $L_{Th} \lesssim \xi$ and well satisfied for $L_{Th} \gg \xi$ in the metallic phase, as expected.

In the insulating or strongly localized case with $L_{Th} \gg \xi$, one *cannot* identify $g(L_{Th})$ with the physical conductance. We first use the result in Eq. (3.5) to obtain the conductivity on scale ξ. Since $g(\xi)$ is a numerical constant, we find that $\sigma_{KG}(\xi) \sim (e^2/\hbar)\xi^{2-d}$, assuming that here the scaling determines the macroscopic conductivity, and using Eq. (3.5) and the definition of w, we obtain

$$\sigma(T) \sim e^2 n(0)\xi^2/\tau_{in} \ . \tag{3.15}$$

This has been obtained previously, using different arguments, by Thouless (1977) and Gogolin and Zimanyi (1983). Equation (3.15) is also in agreement with recent experiments (Ovadyahu and Imry, 1985).

We now turn to discuss the implications of the condition in Eq. (3.8) to our mesoscopic systems. Since we are specifically interested in the regime $L \ll L_{Th}$, in order to have strong interference effects, the LHS of Eq. (3.13) is much larger than unity, thus Eq. (3.13) is *never* satisfied near the localization transition (or weak to strong localization crossover at $d \lesssim 2$). Moreover, even when the sample is only lightly disordered and $V/w \gg 1$, there will always exist low enough temperatures where Eq. (3.13) will be violated. In all these cases the conductance of the sample will vanish as $T \to 0$ according to Eq. (3.5) (we reiterate that this is the conductance defined by absorption, say in a cavity, not the Landauer-type conductance, which will be discussed in the next section). Hence, there are three regimes in terms of the ratio L/ξ. For $L/\xi \gg 1$, in the insulating phase, we come back to the bulk result, Eq. (3.15). Perhaps surprisingly, the same result, except for a numerical factor, also holds for $L \gg \xi$ in the metallic phase. The different numerical factors are simply the values of $g(\xi)$ in the two cases. For $L/\xi \ll 1$, in both metal and insulator at $d > 2$, σ is given near the transition by

$$\sigma(T) \cong g_c(\xi/L)^{2-d} e^2 n(0)L^2/\tau_{in} \quad , \tag{3.15'}$$

while for a lightly disordered $(L/\xi) \ll 1$ system at $d \leqslant 2$ (where $g(L) \gg 1$)

$$\sigma(T) = g(L)e^2 n(0)L^2/\tau_{in} \quad . \tag{3.16}$$

All three relationships, Eqs. (3.15), (3.15') and (3.16) are directly amenable to experimental investigation.

4. The Landauer-type Formulation for Conductance in a Mesoscopic System and Some of its Applications

4.1. Introduction

This formulation is especially suited to discuss the conductance of a segment of a (possibly disordered) system to which two appropriate contacts are made. In 1957 Landauer first introduced the 1D version, which consisted of a given barrier connected with ideal 1D wires (flat potentials) to some external source which drives a current I through the 1D system. The barrier is characterized by its transmission coefficient T and reflection coefficient $R = 1 - T$ (for linear transport and at zero temperature, we need T and R at the Fermi energy). Landauer first considered neutral particles and obtained the density difference across the barrier, thence the appropriate diffusion coefficient. Then the Einstein relation was invoked to obtain the conductance. For charged particles, self-consistent screening (Landauer, 1957) yields the same result. The conductance *due to the barrier* is given by Landauer (1957, 1970)

$$G = \frac{e^2}{\pi \hbar} \frac{T}{R} . \tag{4.1}$$

We emphasize that this is the conductance of the barrier *itself,* defined as the ratio of the current, I, through it and the electrochemical potential difference *across* it. Some confusion has been generated (and later clarified) in the literature, due to the following circumstance. A common way to drive a current through the system is to connect the ideal wires on its two sides to particle reservoirs of chemical potentials μ_1 and μ_2 ($\mu_1 > \mu_2$) as in Fig. 5. If one now computes a conductance G_c using the ratio of I and $\mu_1 - \mu_2$ one obtains

$$G_c = \frac{I}{\mu_1 - \mu_2} = \frac{e^2}{\pi\hbar}T \quad , \tag{4.2}$$

while the previously defined G is given by $G = I/(\mu_A - \mu_B)$, where μ_A and μ_B are the chemical potentials on the ideal wires on the LHS and RHS of the barrier (See Fig. 5). Thus, G_c (which turns out to be smaller than G) is the conductance measured between the two outside reservoirs. Even for $T = 1$, $G_c = (e^2/\pi\hbar)$ is the finite conductance (Büttiker *et al.*, 1985) due to the narrow channel between the two large reservoirs (Sharvin, 1965; Jansen *et al.*, 1983), which can be thought of as due to the two contact resistances, $(\pi\hbar/2e^2)$ each, between the wires and the corresponding reservoirs. In fact, $G_c^{-1} = G^{-1} + (\pi\hbar/e^2)$, i.e., the total resistance between the reservoirs is the sum of the barrier resistance and the two contact resistances.

These contact resistances are due to the geometry of a narrow channel feeding into a large reservoir and to the electrons thermalizing in the baths by inelastic scattering. The analogous contact resistance per channel will turn out to be of the same order of magnitude in the multichannel case, as will be discussed later. It thus becomes an interesting issue as to whether this order

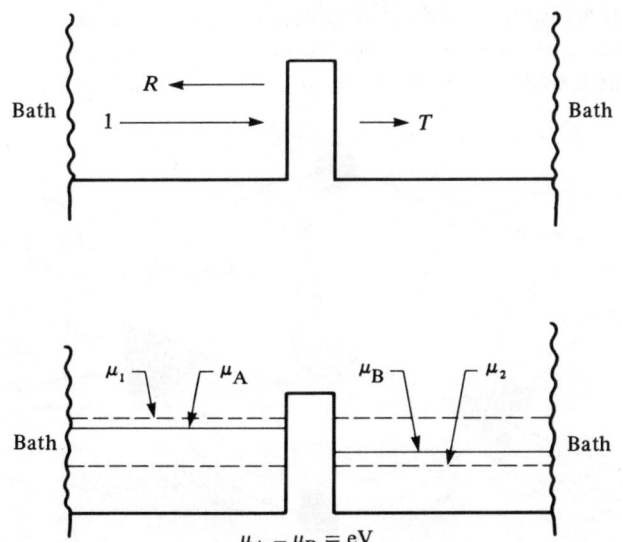

Fig. 5. The Landauer geometry: μ_1, μ_2 are the chemical potentials of the baths; μ_A and μ_B those of the ideal conductors on the two sides of the barrier.

of magnitude is a universal quantity (as one would tend to suspect) or whether it depends strongly on the details of the connection of the thin wire to the reservoir.

The conceptually correct way to measure the chemical potential difference $\mu_A - \mu_B$ was suggested by Engquist and Anderson (1981) and further discussed by Büttiker *et al.* (1985) and Sivan and Imry (1986a). It will be discussed in some details later in this article.

Since the "barrier" may be, in the 1D case, any segment of a linear chain, the above can be used to consider the conductance of any 1D problem. In fact, very early (1970), Landauer has obtained the addition law of two such barriers or "quantum resistors" in series and thence, by induction, the resistance of a linear chain with n randomly placed barriers. The exponential increase of this resistance with n for n larger than some characteristic size (the localization length, in modern terminology) was obtained. This has been the first demonstration of 1D localization in the resistance and the basis for the scaling theory for localization in 1D, which was presented by Anderson *et al.* (1980), after identifying the appropriate variables to be averaged. One should also note that the resistance of two such resistors in series is typically larger than that given by the usual $R_1 + R_2$ law.

The case of two parallel 1D resistors using the Landauer formulation has been first solved by Gefen, Imry and Azbel (1984a) who also found that the addition law is *different* from the classical one. By introducing Aharonov-Bohm type flux Φ in the space inside the loop formed by the two resistors (see Fig. 6) they obtained oscillations in the transmission coefficient and hence of the

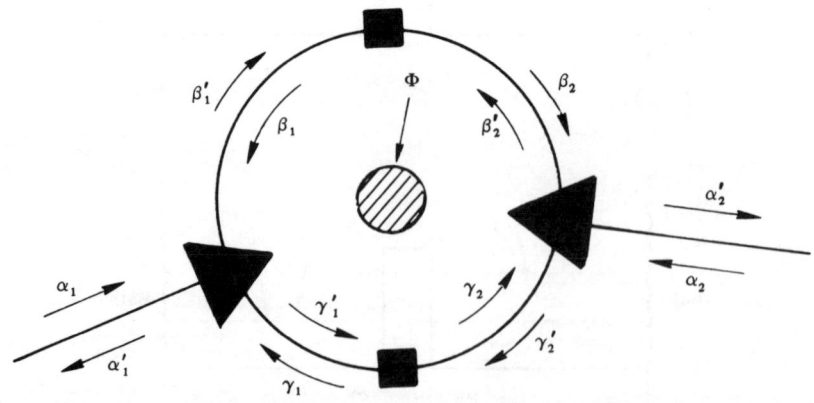

Fig. 6. The model used by Gefen *et al.* (1984a, b): ideal leads are joined by splitters to a 1D ring with two scatterers. A flux Φ is applied through the ring's opening.

resistance between the two 1D leads as function of Φ with a fundamental, usually dominant (Gefen *et al.*, 1984b) period of Φ_0, in agreement with the general expectations discussed in the previous sections.

The generalization of the Landauer approach to the multichannel case is of interest in order to, for example, create a scaling theory for localization in more than one dimension (Anderson, 1981). In this paper we shall be mainly interested in the application of this formulation to the resistance of either a small piece of wire or a small ring-type structure, and in particular, the sensitivity of those to magnetic fields. We shall, thus, first review the multichannel conductance formulas (Anderson *et al.*, 1980; Azbel, 1981; Anderson, 1981; Fisher and Lee, 1981; Langreth and Abrahams, 1981; Büttiker *et al.*, 1985), then discuss the application to the ring with Aharonov-Bohm flux, then the problem of the small wire. We shall also include in this section a brief discussion of the thermoelectric transport (Sivan and Imry, 1986a). One amusing feature of these results is that the Onsager-type relationships such as $G(H) = G(-H)$, or the appropriate equality between two thermoelectric coefficients may not hold (Büttiker and Imry, 1985; Sivan and Imry, 1986a) for the coefficients, such as G, defined on the system *itself*. They do, however, hold for the coefficients, such as G_c, defined between the outside reservoirs. Only in the latter case is inelastic scattering and hence energy dissipation — which appears in an essential fashion in the Onsager derivation — included.

We also point out that this formulation is applicable to many other problems such as the scanning-tunneling-microscopic (Binning *et al.*, 1982) as well as various interface resistances (Castaing and Nozieres, 1980; Uwaha and Nozieres, 1985). Many generalizations, beyond straightforward ones to, e.g., phonon transport, are possible. For example, the inclusion of inelastic processes in the system itself (Büttiker, 1986a; Buttiker and Imry, 1986) should be discussed. The generalization to finite frequencies is a very important open problem, so is a Landauer-type theory for the Hall effect (Büttiker, 1985b; Entin-Wohlman *et al.*, 1986).

4.2. The multichannel Landauer formulation

We now consider the multichannel and finite temperature generalization of the Landauer formulation, depicted in Fig. 7. The leads feeding into the general elastic scattering system S are now ideal wires with a finite cross section A. Due to the quantization in the transverse direction leading to discrete transverse energies E_i, we now have $N_\perp$ conducting channels at the Fermi energy E_F, each characterized at zero temperature by a longitudinal wave vector k_i (and velocity,

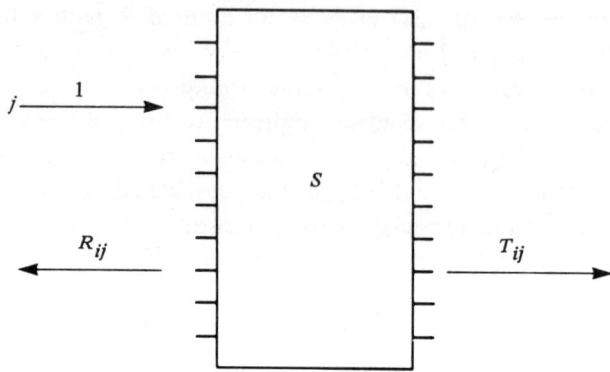

Fig. 7. A multichannel scatterer S. An incoming wave in channel j from the left with amplitude 1 has probabilities R_{ij} and T_{ij} to be reflected or transmitted into the ith LHS or RHS channel, respectively.

$\hbar k_i/m = v_i$), so that

$$E_i + \hbar^2 k_i^2/2m = E_F , \qquad i = 1, \ldots, N_\perp . \tag{4.3}$$

$N_\perp$ is of the order of magnitude of Ak_F^2. At finite temperatures the values of k_i acquire a finite thermal width. The incoming channels (right-going on the LHS and left-going on the RHS) are fed from electron baths with chemical potentials μ_1, μ_2, and the overall temperature is T (the case with $T_1 \neq T_2$ is discussed by Sivan and Imry (1986a) and will be briefly mentioned in this section). We assume that there are no phase relationships among electrons in different channels ("incoherent sources") and consider the case where $\mu_1 - \mu_2$ is small enough to be in the linear transport regime. The system S scatters in the following fashion: An incoming wave (see Fig. 7) from the left jth channel has probabilities $T_{ij} = |t_{ij}|^2$ and $R_{ij} = |r_{ij}|^2$ for transmission into the RHS ith channel and reflection into the LHS ith channel, respectively. The analogous matrices for incoming waves from the RHS are denoted by primes. The $2N_\perp \times 2N_\perp$ matrix S given by

$$S = \begin{pmatrix} r & t' \\ t & r' \end{pmatrix} , \tag{4.4}$$

is unitary due to current conservation, because the T_{ij}, R_{ij} matrices transform the lead *currents*. Furthermore, when time-reversal symmetry holds,

$$SS^* = I ; \qquad S = \tilde{S} , \tag{4.5}$$

where the star denotes complex conjugation, the tilde the matrix transpose and I is the unit matrix.

We define the total transmission and reflection probability into the ith channel by

$$T_i = \sum_j T_{ij} \; ; \qquad R_i = \sum_j R_{ij} \; , \tag{4.6}$$

with similar definitions for the printed quantities. The unitarity conditions imply

$$\sum_i T_i = \sum_i (1 - R_i) \tag{4.7}$$

with a similar condition for the primed quantities.

We also note, for completeness, that the more detailed equalities

$$R_i' + T_i = 1 \; , \qquad R_i + T_i' = 1 \tag{4.8}$$

are valid only between transmission *to* the right (left) and reflection *from* the right (left). They also state that if all incident channels on both sides of the barrier are fully occupied, all outgoing channels will also be fully occupied.

Further unitarity conditions follow from the columns of S. There are additional conditions (see Eq. 4.5) that exist when time-reversal symmetry holds.

As long as they satisfy the above constraints, the elements of S can otherwise be completely arbitrary and they, in principle, depend on energy (although this is a small effect at low temperatures). We note that our assumptions on the incoming channels plus the known matrix S determine completely all the distributions of the outgoing channels. While these are rather out of equilibrium — there are no processes that give the usual "shifted Fermi distribution" or even transfer electrons among the channels to make the chemical potentials of the latter equal — everything is known about them. One can straightforwardly calculate all currents, electron densities, energy densities, entropy densities, etc. We shall discuss below how to define and measure the chemical potentials μ_A and μ_B which will immediately give us the effective conductance of the system G. Since there are good methods to compute S for given models (such as a tight-binding Anderson one) this formulation is suitable for numerical computation.

Before presenting briefly the derivation of the conductance, we point out that ours are not the only possible assumptions. For example, Langreth and Abrahams (1981) had assumed that the various channels on the leads reach a common chemical potential on each side, presumably via electron-electron interaction, and then dropped the assumption of Fermi distributions in the input channels.

While we think that this latter assumption is extremely reasonable and is, in fact, analogous to the known distribution of photons coming out of a photon bath (Landau and Lifschitz 1959), we certainly cannot rule out physical situations where the assumptions of Langreth and Abrahams might apply.

Since the densities of states in the channels are 1D-like, and given, including spin, by

$$n_i(E) = (\pi \hbar v_i)^{-1} \ , \tag{4.9}$$

we write the current on the RHS as

$$I = \frac{e}{\pi \hbar} \sum_i \int dE \, [f_1(E) \, T_i(E) + f_2(E) \, R_i'(E) - f_2(E)]$$

$$= \frac{(\mu_1 - \mu_2)e}{\pi \hbar} \int dE \, -\frac{\partial f}{\partial E} \, \sum_i T_i(E) \ , \tag{4.10}$$

where the velocities cancelled with the density of states factors, and in order to get the last equality we used Eq. (4.7) and stayed in the linear transport regime. It is straightforwardly checked that the current on the LHS is equal to I (conservation of current). The conductance measured between the outside reservoirs, similarly to Eq. (4.2), is thus given by

$$G_c \equiv \frac{I}{\mu_1 - \mu_2}$$

$$= \frac{e^2}{\pi \hbar} \int dE \left(-\frac{\partial f}{\partial E}\right) \sum_i T_i(E) \xrightarrow[T \to 0]{} \frac{e^2}{\pi \hbar} \mathrm{tr}\, t t^\dagger \ , \tag{4.11}$$

where at the zero temperature limit everything is evaluated at E_F. The simplest way to arrive at the conductance of the sample *itself*, defined by $I/(\mu_A - \mu_B)$ is to define μ_A and μ_B as follows. The electron density on the LHS is

$$n_l = \frac{1}{2\pi \hbar} \int dE \sum_i \frac{1}{v_i} [(1 + R_i) f_1 + T_i f_2] \ . \tag{4.12}$$

If the electron gas on the LHS were in equilibrium, with a chemical potential μ_A and Fermi function f_A, its density would be

$$n_A = \frac{1}{\pi \hbar} \int dE \sum_i f_A(E)/v_i \; .$$

(4.13)

We define μ_A so that $n_1 = n_A$, which is a very reasonable definition. It also has the advantage that the Einstein relation will be automatically satisfied (actually, the definition of $\mu_A - \mu_B$ is equivalent to the Einstein relation). Using the value of $(\mu_A - \mu_B)$ thus obtained, one arrives at the result

$$G = \frac{I}{\mu_A - \mu_B}$$

$$= \frac{2e^2}{\pi \hbar} \frac{\int dE \frac{\partial f}{\partial E} \sum_i T_i(E)}{\int dE \sum_i \frac{\partial f}{\partial E} [1 + R_i(E) - T_i(E)]/v_i} \xrightarrow{T \to 0} \frac{2e^2}{\pi \hbar} \frac{\sum T_i \sum 1/v_i}{\sum (1 + R_i - T_i)/v_i} \cdots$$

(4.14)

In the important case where all the T_i's are small, $T_i \ll 1$, $1 + R_i \cong 2$ and the difference between G and G_c is small. This should be applicable for large $N_\perp$ whenever $G \ll e^2 N_\perp/\hbar$, or sample length $L \gg l$. This clearly happens near the localization transition where $G \sim e^2/\pi \hbar$. The zero temperature limit has been first obtained by Azbel (1981) from apparently similar assumptions. The derivation was later clarified and substantiated by Büttiker *et al.* (1985). The generalization of finite temperatures has been considered by Büttiker (Büttiker, 1985a) and by Sivan and Imry (Sivan and Imry, 1986a) who also discussed a subtlety with the definition of G, to be mentioned later on. Even the zero temperature limit in Eq. (4.14) does not agree (except in the important case where the transmissions are small enough for G_c to be a good approximation to G) with all the other multichannel results which existed previously in the literature (Anderson *et al.*, 1980; Abrahams *et al.*, 1979; Fisher and Lee, 1981; Anderson, 1981; Langreth and Abrahams, 1981). This can be easily appreciated by noting that even in the simple independent channel case, Eq. (4.14) does not reduce to the parallel addition form [which will look like $\sum_i T_i(1 + R_i - T_i)^{-1}$]. The reason is, of course, that the latter also assumes a common electrochemical potential difference for all channels. The disagreement with Langreth and Abrahams (1981) has already been discussed above. Discussions overlapping the one here and in Azbel (1981) existed previously in the context of the problem of Kapitza resistance (Castaing and Nozieres, 1980; Uwaha and Nozieres, 1985). It is also important to note that Eq. (4.14) is similar to the finite temperature single-channel result of Engquist and Anderson (1981).

The latter can be obtained, with a small modification to be mentioned later, even from the zero temperature version of Eq. (4.14) by regarding the different energies as (a continuum of) independent channels. It is indeed not of the parallel addition form (for a caveat, see Sivan and Imry (1986a)).

Engquist and Anderson (1981) also introduced the conceptual method by which the chemical potentials μ_A and μ_B could be measured. This is accomplished in principle by bringing in two "measurement reservoirs" with temperature T and with adjustable chemical potentials μ'_A and μ'_B. They are now allowed to (weakly, in order not to disturb the system!) exchange electrons with the LHS and RHS wires, respectively. Now, one adjusts, e.g., μ'_A until no *net* current flows between the appropriate measurement reservoir and the LHS wire. Then, one applies the same procedure to μ'_B. By very general principles it is at this point where the chemical potentials of the measurement reservoirs equal those of the measured systems

$$\mu'_A = \mu_A , \qquad \mu'_B = \mu_B .$$

This clearly accomplishes a four-terminal measurement. The voltmeter (measuring $\mu_A - \mu_B$) does not draw any current.

Three subtleties have to be handled here. First, as already mentioned, the coupling of the measurement reservoir to the system has to be weak enough. Otherwise, although drawing no *total* current it might effect some interchannel electron transfer, by drawing current from some channels and sending it into others (Castaing and Nozieres, 1985). Note that this effect (which has to be avoided in the conceptually correct experiment, but might well exist in a given experimental system) contributes therefore towards interchannel equilibration. (Thus, in a particular, in principle, undesirable limit where the coupling is strong enough for the measurement process to affect the chemical potential of the system, one might expect to get a physical situation similar to the one assumed by Langreth and Abrahams (1981).) Such effects were recently considered by Büttiker (1986a), who, however, finds a tr $tt^\dagger$ -type result when the coupling to the measurement reservoir is strong. It is an interesting and an unavoidable characteristic of our mesoscopic system that normally irrelevant details of the measurement process, may, thus, afftect the results.

Second, we have to understand the requirement on the energy dependence of the coupling between the measurement reservoir and the system. Obviously, we do not want the measurement to distinguish among right- or left-moving electrons, different channels, or energies. To obtain the detailed condition on the coupling, we take the reservoir to have a density of states $n_r(E)$ and to be coupled to the jth channel of system with matrix elements $V_j(E)$. The condition of zero net current from the reservoir μ_A to the system, using the

Fermi golden rule, reads

$$\int dE \sum_i n_r(E) \, | \, V_i(E) \, |^2 f_A(E) \, [2 - f_1(1 + R_i(E)) - f_2 T_i'(E)] \, / v_i$$

$$= \int dE \sum_i \frac{1}{v_i} [f_1(1 + R_i) + f_2 T_i'] \, n_r(E) \, (1 - f_A(E)) \, | \, V_i(E) \, |^2 \quad , \qquad (4.15)$$

where the RHS gives the current from the reservoir to the system as the integral of the density of available electrons times the absolute value squared of the matrix elements, times the final density of states times the density of available holes in the system and the RHS similarly yields the current from the system to the reservoir. Using the relationship in Eq. (4.8) and comparing the relationship in Eq. (4.15) with the one obtained by equating Eqs. (4.12) and (4.13), we find after some algebra that the two definitions of μ_A are equivalent, provided $n_r(E) \, | \, V_i(E) \, |^2$ is *independent of E and of the channel number i* (Sivan and Imry, 1986a). It might be argued that this is a nontrivial condition to satisfy exactly. For a large number of channels, however, the important requirement is that there be no *systematic* variation of $n_r(E) \, | \, V_i(E) \, |^2$ with either E or i. The effects of such variations that are random will tend to average out. Thus, it stands to reason that a real measurement may qualify in this respect and give an unbiased determination of $\mu_A - \mu_B$.

Third, the quantity G evaluated here is an *effective conductance* in the sense of being the ratio of the measured current to the measured voltage. At finite temperatures, however, it was shown (Sivan and Imry, 1986a) that the above G may also have a (usually small) thermoelectric component. The reason for this is that even when the temperatures of the outside reservoirs, T_1 and T_2, are equal, the temperatures that are measured on the two sides of the sample, T_A and T_B (analogous to μ_A and μ_B) will in general be different, at a finite overall temperature. The current I may thus have a component due to the nonvanishing of $T_A - T_B$. This observation is relevant only when the denominator of the conductance formula (Eq. (4.14)) is important (i.e., when the approximation $G \cong G_c$ is not valid). The condition for G_c to be a good approximation to G is that $G \ll N_\perp e^2 / \hbar$, which should be equivalent to the sample length L being much longer than the elastic mean free path l.

4.3. On the question of the contact resistances

As in the case of the single channel, there exist "contact resistances" between the ideal narrow wires and the massive reservoirs. To obtain these, let us consider the case where the whole system S is an ideal conducting wire, i.e., $T_{ij} = T_{ij}' = \delta_{ij}$, $R_{ij} = R_{ij}' = 0$. In this case, $G_c = N_\perp (e^2 / \pi \hbar)$ and the resistance per contact per channel is $(\pi \hbar / 2e^2)$, as in the single channel case. The situation here is identical to that of a small narrow orifice (i.e., a "point contact" (Jansen *et al.*, 1983)) between two large conductors. The resistance of such a contact, with a cross section A, such that $A \ll l^2$, where l is the mean free path in the conductors (the reservoirs in our case) has been calculated by Sharvin (1965). His result is

$$R_{\text{orifice}} = 4\rho l/3A \ , \tag{4.16}$$

where ρ is the resistivity of the conductor. We note that this quasiclassical kinetic theory result is independent of l. Since the number of channels in a square orifice of area A is $A k_F^2/2\pi^2$, the resistance per channel for each contact of the orifice with the conductor becomes (using $\rho = 3\pi^2 \hbar/e^2 k_F^2 l$)

$$R_{c,1\ \text{channel}} = \hbar/e^2 \ , \tag{4.17}$$

which is of the same order of magnitude as the contact resistance alluded to above. We emphasize that this contact resistance has nothing to do with the resistance of the system itself and it exists also when the conductor connecting the two reservoirs is ideal (no scattering). It is just due to the geometry, the ideal orifice being a "bottleneck" between the two conductors, in each of which the electrons are in equilibrium.

It is actually instructive to briefly repeat the Sharvin-type calculation for the resistance of an ideal conductor of cross section A ($= 2\pi^2 N_\perp/k_F^2$) between the "source" and "sink" reservoirs. We take the channel number $N_\perp$ to be much larger than unity to avoid diffraction effects of electrons while entering the ideal conductor. At the same time we also take $A \ll l^2$ (thus the reservoir material should not be too "dirty"). With electron density n_1 and Fermi velocity v_{F_1} in the source reservoir, the number of electrons entering the ideal conductor with an angle θ to the normal, z, to the interface per unit time, is $(n_1/2)Av_z$, where $v_z = v_F \cos \theta$, which yields average (taking spherical Fermi surfaces) currents, $\mathbf{j}_1$, from the source and $\mathbf{j}_2$ from the sink (characterized by the subscripts 1 and 2 respectively):

$$\mathbf{j}_i = (3\pi e n_i/8)A\mathbf{v}_{Fi}/4 \ . \tag{4.18}$$

The net current is due to the difference eV_c in the Fermi energy of the source and the sink, which yields the contact conductance G_{12} between the two reservoirs connected by the ideal conductor with $N_\perp$ channels

$$G_{12} = \frac{\pi e^2}{4\hbar} N_\perp \ . \tag{4.19}$$

This differs by a numerical factor from the Sharvin result, but does also give the required order of magnitude for the contact resistance per channel. The conductance in the pure quantum case was found above to be $G_{12} = (e^2/\pi\hbar)N_\perp$. The difference in the numerical factor is presumably due to a difference between the angular averaging in the classical ballistic way in which the conductor is fed in Eq. (4.19), vs. the equal channel feeding assumed in the pure quantum case. For the particular case of a single channel 1D reservoir, the same numerical coefficient is obtained in both pictures. In the pure 1D case the difference between the reservoirs and the wire is that inelastic scattering and thermalization occur only in the former, while electrons move ballistically along the latter.

4.4. Apparent deviations from the Onsager-type relationships in a magnetic field; thermoelectricity

One of the interesting aspects of the multichannel formula (Eq. (4.14)) is that the various constraints and symmetries in the general case do *not* guarantee the validity of the naive Onsager-type relationships among transport coefficients (Onsager, 1931; Casimir, 1945). For example, the relation $G(H) = G(-H)$ is not guaranteed to hold (although $G_c(H) = G_c(-H)$ is valid, see below). Büttiker and Imry (1985) constructed specific examples with small numbers of channels where

$$G(H) \neq G(-H) \ . \tag{4.20}$$

Numerical calculations by Stone (1985) on larger size disordered models also produced this asymmetry. Such asymmetry also appears in experiments on mesoscopic systems and there are some indications that it might be related to the above-mentioned theoretical possibility of asymmetry. Although it is also possible that magnetic impurities (see Shtrikman and Thomas (1965)) as well as sample inhomogeneity (Von Klitzing, 1985) cause at least a part of the observed asymmetry.

We believe that the reason that the Onsager relation does not have to hold for G is that G measures mostly the effect of the elastic scattering. The inelastic scattering and the associated entropy production, which plays a central role in the Onsager relationships, occur only in the outside reservoirs. Thus, it is indicated that the Onsager relationships *should* hold for G_c of Eq. (4.11), the "two terminal" conductance between the outside reservoirs. It is, in fact, trivial to check (Büttiker and Imry, 1985) that, indeed

$$G_c(H) = G_c(-H) \ . \tag{4.21}$$

Thus the apparent deviation from the Onsager relation for G simply reflects the special nature of the situation in which the inelastic scattering occurs in a region of space away from where the measurement is actually made.

In a recent preprint, Büttiker (1986b) considered the problem of a four terminal conductance measurement, in a generalization of the multichannel formulation. He finds that averaging upon various appropriate contact configurations restores the Onsager symmetry, in agreement with the general Onsager picture and with experiment (Benoit *et al.*, 1986). It might be thought that, therefore, the asymmetry is due to a mixture of a Hall-like effect. We believe, however, that it is also due to the incomplete thermalization of the electrons. An experiment that should clearly demonstrate this point is one where an asymmetry will be obtained with contacts placed on opposite sides of the sample, *parallel* to the magnetic field direction, which should not involve the Hall effect.

A similar apparent deviation from the Onsager relationship for the electronic thermoelectric coefficients measured across the elastic scatterer was also found to occur by Sivan and Imry (1986a). These coefficients are defined by

$$
\begin{pmatrix} I \\ U \end{pmatrix} = \begin{pmatrix} K_0 & (1/T)K_1 \\ \tilde{K}_1 & (1/T)K_2 \end{pmatrix} \begin{pmatrix} \mu_A - \mu_B \\ T_A - T_B \end{pmatrix} ,
\tag{4.22}
$$

where U is the heat current and T_A, T_B the temperatures measured in the leads on the two sides of the scatterer S. The Onsager relation implies $K_1 = \tilde{K}_1$ in this case. T_A and T_B can be defined, similarly to μ_A and μ_B, by adding a condition, analogous to equating Eqs. (4.12) and (4.13) for the specific entropies (or energies) on the two sides. Alternatively, one may invoke a temperature measurement process by demanding that the entropy currents between the measurement reservoirs and the corresponding ideal conductors also vanish, in analogy with Eq. (4.15). Again, the condition that the two definitions coincide is that the product $|V(E)|^2 n_r(E)$ be independent of energy and channel number (or at least not have systematic variations, for a large number of channels).

It was found, as in the case of $G(H)$, that the coefficients defined with the temperature and chemical potential differences between the outside baths do satisfy the Onsager relations but those across the sample (Eq. (4.22)) do not. It is to be kept in mind that these two sets of coefficients do coincide for a large number of channels and/or weak transmission as discussed before. Exact expressions were obtained for all the coefficients as well as low temperature approximations to them via the Sommerfeld expansion. It turns out that both the Wiedemann-Franz relation, $K_2 = (\pi^2/3)(k_B T)^2 K_0$ and the Onsager relation hold only to lowest ($O(T^2)$) order. $O(T^4)$ corrections exist in both.

A quantity which is commonly measured is the thermopower, given by

$$S = \frac{1}{eT} \frac{K_1}{K_0} .$$ (4.23)

To lowest order one obtains the familiar expression

$$S \cong \frac{\pi^2}{3} k_B T \frac{\partial}{\partial E} [\ln K_0 (T = 0)]_\mu .$$ (4.24)

We remark that Eq. (4.24) is more likely to hold here than in pure metals, if the disorder and boundaries limit the phonon mean free path enough to eliminate phonon drag effects. The more general, very useful and physically motivated expression due to Cutler and Mott (1969)

$$S = (k_B/e) \int_0^\infty \sigma(E) \frac{E - \mu}{k_B T} \left(-\frac{\partial f}{\partial E} \right) dE \bigg/ \int_0^\infty \sigma(E) \left(-\frac{\partial f}{\partial E} \right) dE ,$$ (4.25)

is valid only to lowest order *or* for the coefficients measured between the outside reservoirs. The latter are approximately valid for $L \gg l$, as discussed before. So Eq. (4.25) can be used in many cases, including near the metal-insulator transition at temperatures that are not too high.

4.5. *Application to interference effects in rings*

We now turn to the application of the multichannel formulation to the interference effects in small ring-type structures and to their sensitivity to Aharonov-Bohm fluxes in the rings' opening. As already mentioned in Sec. 3.2, the following dilemma appeared to exist in this case. The work of Altshuler, Aronov and Spivak (AAS) (1981), based on an expansion in the disorder, suggests a periodicity in the Aharonov-Bohm flux Φ with a period $h/2e$ equal to half the single-electron flux quantum $\Phi_0 = h/e$. On the other hand, an exact calculation (Gefen *et al.*, 1984a, b) of the conductance of a purely 1D ring, where both the leads and the two branches of the ring are single channel only, yields a fundamental periodicity with a period Φ_0. The amplitude of the Φ_0-periodic oscillation is, in general, larger than that of the first harmonic ($\Phi_0/2$-periodic) oscillation, except for very special cases (Gefen *et al.*, 1984b). It is possible, however, that this circumstance is an artifact of the purely single channel case and that when the number of channels becomes large, the $\Phi_0/2$-periodic term will become dominant. A more specific reason for this expectation is based on the following consideration. The AAS $\Phi_0/2$-periodic oscillation is associated (Altshuler *et al.*, 1982b; Sharvin and Sharvin, 1981; Bergmann, 1984) with a coherent back-scattering due to the constructive interference (for $\Phi = 0$) between two time-reversed paths of waves going in opposite senses around the ring as illustrated in Fig. 2.5 of Bergmann (1984). These waves have, for finite Φ, a phase difference of $4\pi\Phi/\Phi_0$, which leads to a periodicity with a period $\Phi_0/2$, and with a minimum (neglecting spin-orbit effects) transmission for $\Phi = 0$. On the other hand, the lowest order Φ_0-periodic contribution is due to direct interference between two waves propagating along the two branches of the ring, whose Φ-dependent phase difference is $2\pi\Phi/\Phi_0$. Since all the other phases are effectively random, this term may be maximal or minimal at $\Phi = 0$ and it is suggested, therefore, that for many channels this contribution may average out and the $\Phi_0/2$-periodic one may become dominant.

It is easy to use our multichannel conductance formulas to estimate the channel number (N) dependence of those two contributions (Büttiker *et al.*, 1985). We find that both scales go like $1/N_\perp$, so that their ratio stays finite as $N_\perp \to \infty$. Here $N_\perp$ is the number of channels and our model is schematized in Fig. 8. $N_\perp$ incoming channels, i, are split (Shapiro, 1983; Büttiker *et al.*, 1985) into $N_\perp$ in the upper part of the ring (and picking up, for a symmetric case, a phase of $\pi\Phi/\Phi_0$ each from the flux) and $N_\perp$ in the lower branch (with a flux dependent phase of $\pi\Phi/\Phi_0$). These $2N_\perp$ channels are now combined at the end of the ring, into the $N_\perp$ output channels, j.

As argued by Gefen *et al.* (1984a) and Büttiker *et al.* (1985), each transmission coefficient T_{ij} across the whole system has a relative oscillation due to the direct (Φ_0-periodic) interference effect which is "of order one" (i.e., independent of $N_\perp$). We have $N_\perp^2$ such random terms, so that the relative contribution to Eq. (4.14) which is Φ_0-periodic is $O(1/N_\perp)$, (contrary to a speculation of $O(N_\perp^{-1/2})$ by Gefen *et al.* (1984a)). On the other hand, the $\Phi_0/2$-periodic part, which shows itself, e.g., in the R_i terms in the denominators of Eq. (4.14)

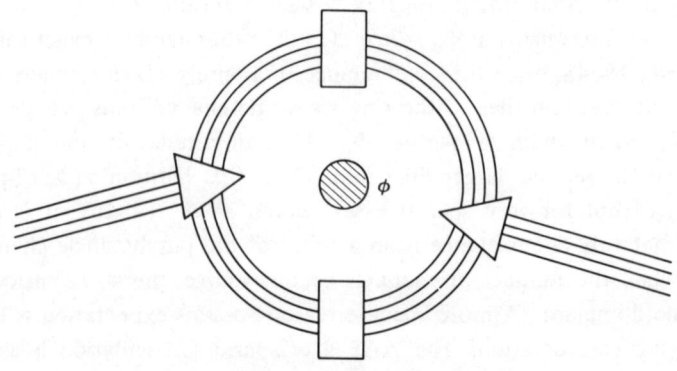

Fig. 8. A multichannel ring with $N = 4$.

is also of relative order $1/N_\perp$. This can be demonstrated by constructing arguments similar to the one sketched above, for the relative size of the part of the conductance with period $\Phi_0/2$. Rather than doing that, we refer the reader to the AAS results. For a thin ring the flux-sensitive quantum correction to the flux insensitive conductance, Eq. (5) of AAS, can be cast in the form

$$\Delta G = (e^2/\hbar) f(2\Phi/\Phi_0) . \tag{4.26}$$

Here f is a periodic function of period one which depends only on the ratio of the circumference to the inelastic length, but not on the cross section of the ring. If we now assume that the flux insensitive conductance is proportional to the cross section, i.e., proportional to the number of channels $N_\perp$, then Eq. (4.26) implies that the relative amplitude of the AAS oscillations with period $\Phi_0/2$ is also of order $1/N_\perp$. Thus, in a particular $N_\perp$-channel ring the terms with two periods under consideration are, *as far as their $N_\perp$-dependences are concerned,* of the same order $(1/N_\perp)$. The dependence on L and I will be discussed in Sec. 5.

We emphasize that the result of AAS, schematized in Eq. (4.26), is a quantitative one for the $\Phi_0/2$-periodic component. On the other hand, our estimate for the Φ_0-periodic one should be regarded only as a lower bound. For example, if correlations among channels exist, or if the contributions of some of the channels are substantially larger than those of the others (due, e.g., to the former having higher longitudinal energies, or being close to a resonance; Lifschitz and Kirpichnikov, 1979; Azbel, 1983), the effective $N_\perp$ in Eq. (4.14) might be *smaller* than the total one. A mechanism for reducing the effective $N_\perp$ by a length-dependent factor will be presented in Sec. 5. Also, as suggested by Büttiker (1985a) the denominator in the multichannel formula (Eq. (4.14)) may well increase the coefficient of $1/N_\perp$ in the Φ_0-periodic term. It, thus, should not have come as a surprise that when the appropriate experiment, as discussed below, did produce the Φ_0-periodic contribution, its strength (measured, e.g., by the size of the appropriate Fourier coefficient) did turn out to be substantially larger than the $\Phi_0/2$-periodic contribution. It should also be kept in mind that since the path length for the latter process is twice that for the former, the $\Phi_0/2$ contribution should thus be more sensitive to inelastic scattering — another reason for its being smaller. The stronger sensitivity of the AAS contribution to magnetic fields in the wires will be discussed later on.

4.6. Ensemble and thermal averaging, aperiodic conductance fluctuations

As discussed in Sec. 3-2 the reason why the Φ_0-periodic contribution did not show up in many experiments performed on small-radius cylinders and arrays of small circuits (Sharvin and Sharvin, 1981; Altshuler *et al*., 1982; Ladan and Maurer, 1983; Gordon, 1984; Gijs *et al*., 1984; Pannetier *et al*., 1984, 1985; Bishop *et al*., 1985; Licini, Dolan and Bishop, 1985; Licini, Bishop, Kastner and Melngailis, 1985; Dolan *et al*., 1986) was the effective ensemble averaging involved there (Gefen, 1984; Carini *et al*., 1984; Browne *et al*., 1984; Büttiker *et al*., 1985; Murat *et al*., 1986; Imry and Shiren, 1986; Stone and Imry, 1986). What we have found here is that this averaging is *not* automatically provided by having many channels. The reason that the $\Phi_0/2$ term survives ensemble averaging is that, as discussed above, this term is constrained (in each ensemble member) to have a maximum resistance at $\Phi = 0$ (we discuss here only the case with no spin-orbit scattering). Since the Φ_0-periodic term does *not* have such a definite "phase of oscillation," it is highly likely to average out over an ensemble of different microscopic systems prepared under the same macroscopic constraints. To demonstrate this property, we show in Fig. 9 (taken from Stone and Imry (1986)) for the many-channel case, the dimensionless conductance g for a given ring with several energies similar to choices of the phases of the individual scatterers on the arms (see below) — demonstrating that g can be maximal or minimal at $\Phi = 0$. If different members of the ensemble have the same impurities in the two arms but are situated at different points along the arms or slightly different lengths of the ideal arms (where the scale is determined by $k_F^{-1} \sim$ a few A!) then it is reasonable to expect the phase of the scatterers to vary almost randomly among ensemble members. Thus, in the right-hand part of that figure we show the ensemble averaged $g(\Phi)$ which in fact has now a $\Phi_0/2$ fundamental period, with a minimum at the origin, as expected. The last figure also shows the equivalence of energy averaging and ensemble averaging; see below.

We thus believe that in order to observe the (possibly relatively large) Φ_0-periodic component, one should use a single, specific ring. Measurements involving effective ensemble averaging will yield a $\Phi_0/2$ fundamental period, as the ensemble-averaged theoretical perturbative calculations do. Recently, it has been shown by Browne and Nagel (1985) that the ensemble-averaged G is $\Phi_0/2$-periodic also in the presence of electron-electron interaction.

The above discussion of the phases of the scatterers suggests another important consideration. These phases are functions of the electron's energy. They will be well defined at $T = 0$. However, at high enough temperatures, different electrons in the range of $k_B T$ around E_F will have "optical paths" around the ring that may be different enough to yield the same effect as ensemble averaging. This effect is again illustrated by numerical calculations in the 1D ring (taken from Murat *et al*. (1986)) in Fig. 10 where at low T one gets a fundamental

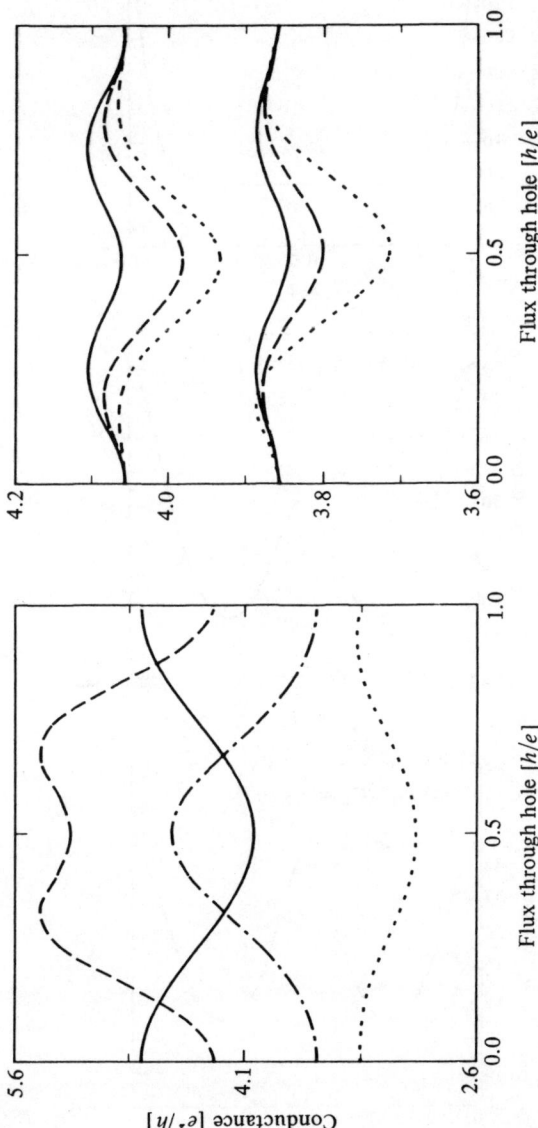

Fig. 9. From Stone and Imry (1986). Left: conductance vs. flux for several energies. Right: comparison of ensemble and energy-averaging, demonstrating transition to $\Phi_0/2$ periodicity.

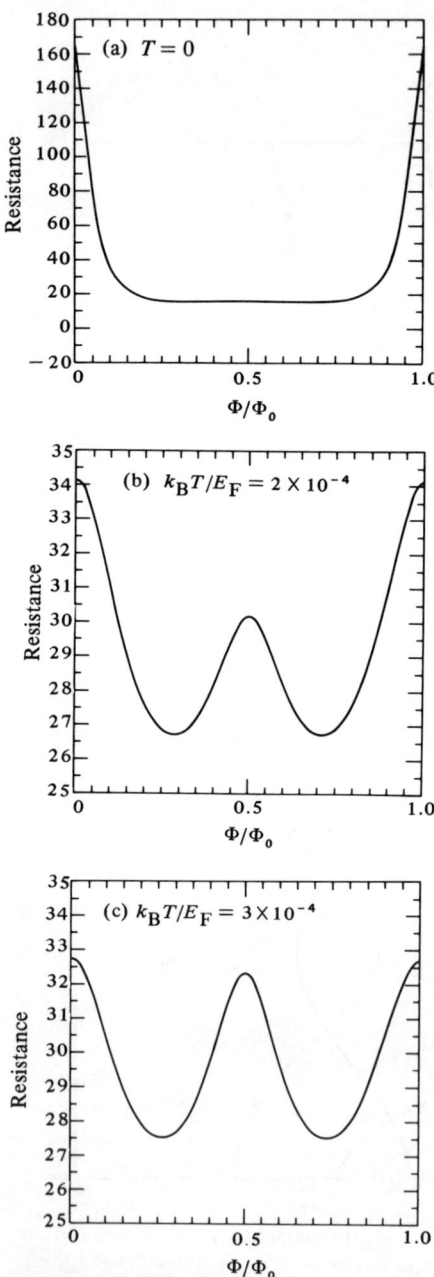

Fig. 10. Similar to Figs. 4 and 5 of Murat *et al.* (1986). Resistance as function of flux for a 1D ring with long arms with scatterers, for three temperatures, exhibiting the self-averaging-out of the Φ_0 periodic component at higher temperatures.

periodicity of Φ_0, but at higher temperatures the basic period reverts to $\Phi_0/2$. The cross-over is obtained, in 1D, when $k_B T$ becomes comparable to the level separation, w, as discussed for the Kubo conductivity in Sec. 3. This should be expected since consecutive levels in 1D are defined by having $O(2\pi)$ more phase variation along the whole system.

More generally, there will exist some characteristic energy scale Δ, so that energies differing by Δ will have significant differences in the interference along the system. In 1D, $\Delta \sim w$. An important question is what is the "energy correlation range" Δ in more general cases. Δ is in fact generally of the same nature as the Thouless parameter V which measures, as discussed in Sec. 3, the sensitivity of the energy levels to boundary conditions (i.e., a phase difference across the system, or an Aharonov-Bohm type flux through the ring). One would thus heuristically identify the "energy correlation range" Δ with V. Another heuristic argument, relying on the longer optical path for diffusion compared with free motion, for this identification is given in Stone and Imry (1986) who also demonstrated the results numerically for the multichannel case. This is also consistent with the results of Lee and Stone (1985) in the weakly localized regime:

$$\Delta \sim V \qquad (4.27)$$

so that the condition that the thermal, "Fermi" smearing will not wash out interference effects, such as the Φ_0-periodic component in a ring, is

$$k_B T \lesssim V . \qquad (4.28)$$

Using the Thouless relation in Eq. (3.15), the expression $w \sim [n(0)\Omega]^{-1}$, the Einstein relation (Eq. (3.16)) and the usual relation between G and σ, Eq. (4.28) is found to be equivalent to

$$L \lesssim \sqrt{D\hbar/k_B T} \equiv L_T \qquad (4.29)$$

i.e., that the sample is shorter than the thermal coherence length defined by the RHS of Eq. (4.29). Comparing this condition with the one stating that L be smaller than the inelastic diffusion length (3.20), and noting that τ_{in} is larger by one to two orders of magnitude than $\hbar/k_B T$ in many real systems (especially in metals, they appear to have similar orders of magnitude in semiconducting systems), we find that our condition in Eq. (4.29) is more restrictive, but not hopelessly severe. Especially because for $k_B T \gg \Delta$, averaging may be expected to reduce the relative oscillation by $\sim (\Delta/k_B T)^{1/2}$. For example, for a $500\text{A} \times 1500\text{A}$ gold wire with a length of 5000A and a resistance of 20Ω, $w \sim 1\,\text{mK}$

and $\mathcal{V} \sim .05\,$K. Hence the condition in Eq. (4.29) is experimentally feasible and the effect should be reduced by only a factor of 3 at $T \sim .5\,$K by energy averaging, while were Δ given by w, the situation would have been much more difficult. This is the range of parameters relevant for some recent experiments.

Until now we discussed only the effect of the Aharonov-Bohm type flux through the opening of the ring. One might also enquire about the effect of the magnetic field in the material. In fact, the initial experiments on rings, where a substantial fraction of the magnetic flux went through their arms, as well as similar experiments on singly connected fine lines showed aperiodic fluctuations (Umbach *et al.*, 1984) in the resistance of these systems, where the random-like structure was reproducible for a given system as long as it was not effectively annealed. The magnetic field scale of this structure corresponds to the flux through the wire, being on the order of a flux quantum. This is reminiscent of the oscillations found by Dingle (1952) in the equilibrium properties of free electrons in the geometry of, say, a disc, as function of the flux through it. One might imagine (Blonder, 1984) that the aperiodic nature of the resistance change in the real system might be due to the random specific stacking of impurities and defects in a given system, where each given stacking should produce as its own fingerprint a specific $R(H)$ curve.

The multichannel conduction formula provides an ideal tool to quantitatively check the above idea. One can model the system by, for example, a disordered tight binding Anderson model. It is possible to calculate the S matrix for a given model numerically either by multiplying the transfer matrices (Pichard and Sarma, 1981, Azbel, 1982) (and variation thereof), or by the Green's function method of Thouless and Kirkpatrick (1981), generalized by 2D by Fisher and Lee (1981). It is reasonable to reach with the latter method, whose convergence properties appear to be better for the cases of interest here, 2D models up to $\sim 40 \times 400$ sites. These calculations have been performed by Stone (1985). The effect of the magnetic field was taken into account by modifying the phases of the transfer matrix elements so that the sum of all phases around each loop is given by the flux inside it ("Aharonov-Bohm effect in each loop"). Typical results (Stone, 1985), where the computer experiments and real experiment have the field scales determined by the condition of a flux quantum through the system are depicted in Fig. 11. The large difference in the vertical scale, where the fluctuations in the computer experiments are larger by more than two orders of magnitude than in the real experiments, is due mainly (see Sec. 5) to the different channel numbers (about forty in the former, 2×10^{4} in the latter). This and many other similar results obtained by Stone (1985) provide convincing evidence that the physics of the aperiodic reproducible oscillations is indeed the modification of the electron interference by the

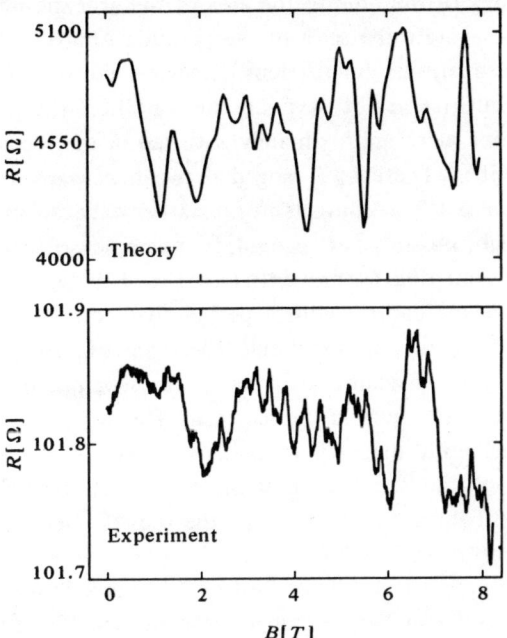

Fig. 11. Resistance as function of flux for a small wire. Theory and experiment normalized to have similar horizontal scales. Vertical scales are discussed in the text. From Stone (1985).

magnetic flux through the system. The qualitative picture is as follows: Each t_{ij}, for example, can be obtained as a sum over all paths through the sample of the transmission amplitude from i to j through the given path, which is clearly an interference phenomenon. A magnetic field having a flux through the system on the order of Φ_0, changes the relative phases of the most distant paths by the order of 2π. This determines the field scale on which the resistance may fluctuate. These ideas are more fully discussed by Stone (1985) and Stone and Imry (1986). The concept of the energy correlation range, Δ, alluded to above for the electrons, emerges very clearly from these numerical results. More general ideas (Altshuler, 1985; Lee and Stone, 1985; Altshuler and Khmel'nitskii, 1985; Imry, 1986), based on Stone's findings about the magnitude of these fluctuations, on conductance fluctuations (Stone, 1985) will be discussed in Sec. 5.

4.7. Conclusions and experiments

We are now in a position to summarize the conditions to observe the h/e periodic oscillation. The experiments should be done on a single specific ring, the ratio of the area of the hole to the area of the arm should be large enough (Stone, 1985) to separate the scale of the periodic Aharonov-Bohm oscillation from that of the aperiodic fluctuations discussed above and the temperature should be low enough so as not to violate the condition in Eq. (4.29) too much. All these conditions were approximately satisfied in the recent experiment by Webb *et al.* (1985b). (There $k_B T$ ranged between perhaps a factor of two to twenty larger than Δ.) The contacts in the earlier experiments had been more massive, which might possibly have caused, for example, some unwanted inelastic scattering. These gold ring experiments (see Fig. 12) did, in fact, show clear and relatively large oscillations with a period that corresponded within $\sim 5\%$ to the flux quantum, Φ_0, in the hole. These are superposed on a structure having a much larger magnetic field range, as *semi-quantitatively* expected (Stone, 1985) due to the field in the material. The structures do appear below ~ 0.5 K, consistent with the estimates made above for the energy averaging. The relative size of the structure is larger than $1/N_\perp$ by a factor of 20–50, which is not a total surprise, remembering the remarks made above concerning the lower limit nature of the $1/N_\perp$ result. This relatively large size is also in keeping with the numerical results of Stone (1985) and Stone and Imry (1986) and the general results in Sec. 5. Finally, we remark that the ratio between the large and small field scales is about 25 while the ratio between the area of the hole and of the arm is about 10. This appears to be consistent with the numerical coefficients in the scale of the aperiodic oscillation, as found by Stone (1985). Soon thereafter, the h/e as well as the AAS oscillations were seen in the same ring by Chandrasekhar *et al.* (1985). Datta *et al.* (1986a) found h/e oscillations in semiconducting multilayer structures.

One result of the experiment of Webb *et al.* (1985b) has been that the Φ_0-periodic oscillations did not appear to fade with increasing magnetic field. They, in fact, persisted to $\sim 10^3$ oscillations with no noticeable weakening. This was quite surprising, due to expectations that the non-Aharonov-Bohm flux inside the arms should make Φ ill-defined and eventually smear the structure, as in fact happens with the AAS-type $\Phi_0/2$-periodic oscillations (Altshuler *et al.*, 1981; Sharvin and Sharvin, 1981; Altshuler *et al.*, 1982b). It is possible, however, at least in the case where the scales of the periodic (Aharonov-Bohm) and aperiodic (due to the field inside the material) structures are well separated, to give a heuristic argument (Stone and Imry, 1986) for the persistence of the periodic structure in the Φ_0-periodic component.

The magnetic field applied to the real ring will create both an Aharonov-Bohm flux Φ through the ring's opening and a (classically relevant) flux Φ_c through the ring's arms. The ratio of these two fluxes is just a geometrical factor

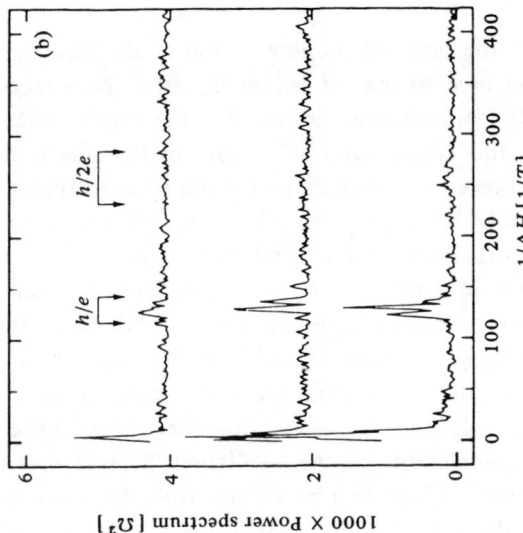

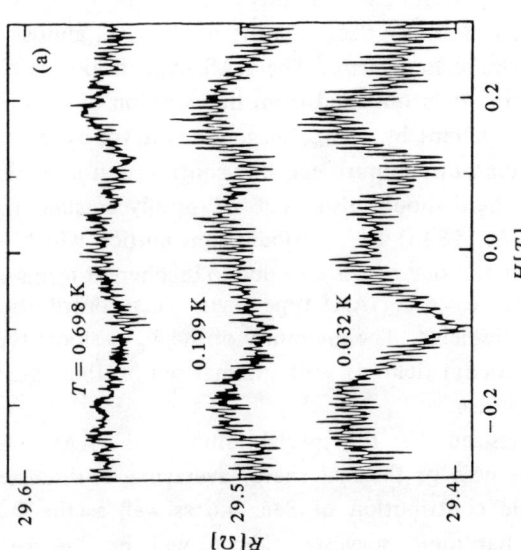

Fig. 12. Resistance oscillations in a small gold ring, from Webb *et al.* (1985b), exhibiting "fast" h/e oscillation over a slow background, due to the field in the wire. The Fourier spectra are also shown.

$$\frac{\Phi}{\Phi_c} = A \ .$$ (4.30)

For H perpendicular to the plane of the ring, A will be the aspect ratio — the ratio of the area of the hole to that of one of the arms. Now recall, that by the results of Stone (1985), mentioned in Sec. 4.6, the scale of changes of Φ_c which markedly alters the conductance of an arm of the ring is $\Delta\Phi_c \sim \Phi_0$. Thus, for $A \gg 1$ it follows that when H is changed so as to span a range Φ_0 of Φ, Φ_c will change only by Φ_0/A. This will cause just a very small change of the background contribution to the conductance due to Φ_c, which will result in a slow variation of $G(H)$ on top of which the faster oscillation due to Φ will occur. Further analysis (Stone and Imry, 1986) along the lines of Eqs. (5.1) to (5.3) of Büttiker *et al.* (1985), with the introduction of the effects of Φ_c indeed reveals that the Φ_0-periodic oscillation survives the existence of Φ_c even when $\Phi_c \gg \Phi_0$, although Φ_c causes slow (determined by $\Delta\Phi_c \sim \Phi_0$) amplitude and phase modulations of the oscillation. This is also consistent with numerical simulations, shown in Fig. 13 and with the weak localization calculation (Lee and Stone, 1985) as well as with the experiment (Webb *et al.*, 1985b) (see Fig. 12).

In essence, the reason that the Φ_0-periodic oscillation survives the large magnetic fields in the ring's arms, is that this contribution is (Büttiker *et al.*, 1985; Stone and Imry, 1986) (see Sec. 4.6) the result of an almost cancelling addition of many random-phased terms. The AAS contribution is of a fundamentally different nature. It is obtained from the addition of coherent terms. Once these are made incoherent by a large magnetic field (or by, say, a random magnetic field due to magnetic impurities), this contribution is further greatly reduced. Interestingly, there should also exist a typically smaller, (Stone and Imry, 1986; Gefen *et al.*, 1984b) $\Phi_0/2$-periodic contribution which is the first harmonic of the Φ_0-periodic one and is also due to incoherent terms and *not* to the special coherent-backscattering AAS-type ones. This contribution should also survive large magnetic fields. The immunity of the Φ_0-periodic type oscillation to (uniform or random) fields is well summarized by the statement that "you cannot kill a dead horse."

It should be reemphasized that the special nature of the AAS contribution makes it withstand ensemble or thermal energy averaging, as discussed in Sec. 4.5, while the aperiodic contribution of Sec. 4.6 as well as the Φ_0-periodic contribution and its harmonic suggested above will be averaged out. It is interesting how these processes have different sensitivities to the various parameters and each is observable under different experimental conditions. A final remark is in order about the strengths of these phenomena. At

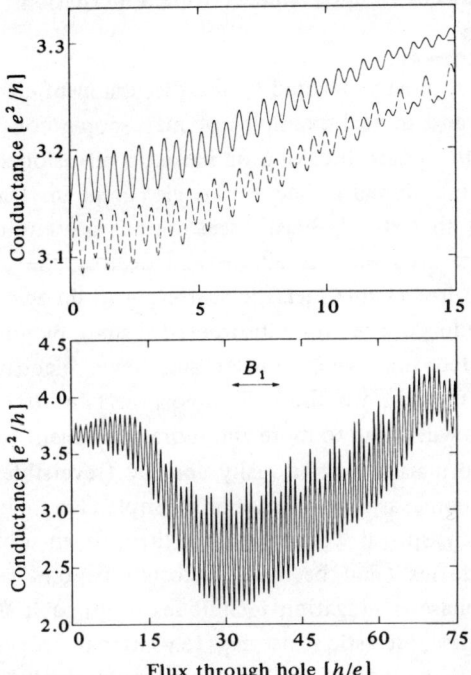

Fig. 13. Numerical simulation of the magnetoconductance of a 100 × 10 site ring with an aspect ratio of 4, from Stone and Imry (1986). Bottom: h/e oscillation in a single ring. B_1 is the predicted scale of aperiodic background. Top: Ensemble-averaged magneto-conductance for 200 (dashed line) and 950 (solid line) samples for the same system, exhibiting the emergence of the $h/2e$ periodicity at small fields. Note the return to h/e periodicity around the scale B_1 for the 200-samples case.

higher temperatures, when $L_\Phi \ll L$ the h/e and AAS oscillations should die out, respectively, like e^{-L/L_Φ} and e^{-2L/L_Φ} (due to the larger path length of the latter contribution where L is the circumference of the ring). On the other hand, for the aperiodic oscillation, coherence does not have to exist along the whole ring. For $w \ll L_\Phi \ll L$ (w being the width of the line) there are basically just L/L_Φ independent fluctuating resistances (Fukuyama *et al.*, 1986; Altshuler and Khmel'nitskii, 1985) so that the relative fluctuation should decrease like $(L_\Phi/L)^{1/2}$ (with further averaging for $L_T \ll L_\Phi$) i.e., *algebraically* in T. Similarly, for n rings in series (Umbach *et al.*, 1986) with $L_\Phi \sim L$ (to make them incoherent) the AAS effect (on a relative scale) will stay independent of n, while the h/e and aperiodic fluctuations should decrease in relative magnitude like $n^{-1/2}$. Their magnitude is also decreased by energy averaging (Murat *et al.*, 1986; Imry and Shiren, 1986; Stone and Imry, 1986) since $L_\Phi \gg L_T$ most of the time.

5. Conclusions, Recent Results, Nonelectronic Fluctuations, Open Problems

This paper has been mostly devoted to the discussion of electronic interference effects in models and in real examples of mesoscopic conductors whose sizes are smaller than the phase breaking or inelastic diffusion length. The energy correlation range, mentioned in Sec. 4 may also come in. One principle that we established here is that strictly elastic scattering is not enough to scramble the phases of the electron's wave functions and destroy the interference effects. It takes effectively irreversible inelastic scattering to do that, and the effects of the latter can be reduced, even for a disordered system, by going to low temperatures. It is expected that when smaller and lower electron density systems become available (which, we think, is amost certain to occur) the relevant temperature ranges will go up to more and more convenient ones.

That phase information is not really lost by (reversible) elastic scattering should not really come as a surprise. For example (Bergman, 1983), it is now well known that an optical wave that goes through an arbitrary complicated set of static "obstacles" and becomes distorted beyond recognition, can be time-reversed by phase conjugation techniques to go back and yield the initial wave form. Of course, inelastic scatterings (e.g. Raman processes, dust particles) may destroy the above possibility. Ingenious pulse methods can be used to recover phase information in magnetic resonance experiments.

The separation between the scales of elastic and inelastic scattering, where the dissipation and equilibration occur away from where the transport coefficients are determined by elastic scattering leads to novel regimes of transport theory where many new phenomena (e.g. apparent deviation from the Onsager-type relationships, interface impedances and, more generally, those due to inhomogeneities, to mention but a few) occur. More generally, the scale for elastic scattering at the appropriate energy can also be made larger than the system's size — such as in the extremely important ballistic devices (Heiblum *et al.*, 1985). These effects should be describable by the fuller conduction formula (Eq. (4.14)). Thus several new regimes due to the interplay among the various scales mentioned above, are possible. We expect many new phenomena to possibly show up in these systems and fundamental theoretical developments, including the ability to treat systems that are further from equilibrium than the usual ones, will be necessary.

While some of the results we presented, such as the exact periodicity as function of Φ, are general and include electron-electron and electron-static lattice interactions, most of our discussion has been in terms of the independent electron (or quasi-particle) picture. Generalizations to include some effects of the interactions (Altshuler and Aronov, 1985) should be possible — for example, the effects of weak electron-electron inelastic scattering or the recent proof (Browne and Nagel, 1985) that ensemble averaging produces a fundamental period of $h/2e$, in the presence of interactions. Clearly, a lot of further work concerning the electron-electron interactions will have to be done. In this paper we did not discuss the complications due to spin-orbit interactions and those with magnetic impurities (except to mention that it is easy to generalize the statements in Sec. 4.7 concerning the different sensitivities of the h/e and AAS oscillations to magnetic fields, to include at least the orbital effect of random magnetic fields due to magnetic impurities). These are very useful experimental handles that can be varied to yield important information on these systems (Altshuler *et al.*, 1982a; Bergmann, 1984; Lee and Ramakrishnan, 1985).

Various further generalizations and extensions of the work reviewed in this article are called for, and some of them are forthcoming. A more detailed understanding of the processes that can cause real phase incoherence is needed. Once these effects exist, how exactly do they modify the phenomena studied here? An interesting way to study the effects of thermalization via coupling to an "incoherent" electron bath, was suggested by Büttiker (1985b). As already mentioned, it is necessary to generalize the Landauer-type treatment to account for inelastic scattering (Büttiker and Imry, 1986) and energy relaxation, finite frequencies and the Hall effect. The latter will be discussed below, its understanding may pave the way towards a microcopic theory of, for example, the quantized Hall effect (Von Klitzing *et al.*, 1980) in disordered systems (Laughlin, 1981), as well as finite size and, hopefully, Josephson-type effects associated with it (Imry, 1983b). It is interesting that one of the strongest theoretical pictures (Laughlin, 1981) for understanding of the (integer) Hall effect employs a cylindrical geometry similar to the one discussed here. Finally, the multichannel generalization as well as experimental observation of the persistent diamagnetic currents and ac Josephson-type effect discussed in Sec. 2 are important open problems.

We now briefly discuss two recent interesting developments relating to the above material, the Hall effect (Büttiker, 1985b; Entin-Wohlman *et al.*, 1986; Nguyen *et al.*, 1985; Holstein, 1961) and universal conductance fluctuations, where the latter may suggest a model for low frequency noise (Dutta and Horn, 1981). We shall then conclude the paper by discussing effects that are not necessarily electronic.

5.1. The Hall-type effect

Some promising developments on the theory of the Hall-type effect in mesoscopic systems have recently occurred. To get a voltage perpendicular to the
current direction, one needs at least two conduction channels. The simplest
such model is the one of Gefen *et al.* (1984a): a 1 D ring coupled to single
channel leads, as depicted in Fig. 6. Taking the leads to be parallel to the x-
direction, and the current to be, thus, in the same direction, the Hall effect is
related to a voltage between the two arms of the ring, in the y-direction. Another
interesting quantity in this respect is the current circulating around the ring due
to the difference in the currents in its upper and lower branches.

Since one can define the chemical potentials on the ideal-conductor portions
of these arms, similarly to the discussion in Sec. 4.2, the Hall voltage can be
straightforwardly calculated. This has been done independently by Büttiker
(1985b) and by Entin-Wohlman *et al.* (1986). The ring is seen to have a
relatively large Hall conductance (including a nonzero G_{xy} at $\Phi = 0$, if the
symmetry is low enough). For a given current the Hall voltage oscillates as a
function of Φ, with a period Φ_0. Thus, although it does not produce a direct
Lorentz force, as in the classical picture, the Aharonov-Bohm flux Φ does yield
a very interesting, oscillating Hall effect.

A Hall effect due to an essentially Aharonov-Bohm flux has already been
first found by Holstein (1961) many years ago in his theory of the Hall effect
in hopping conduction. The magnetic field sensitivity of the hopping probability
between two localized site states is due to a coherent superposition of the direct
transition with an indirect higher order one, involving an intermediate state on a
third site. An interesting feature of this theory, which is not always appreciated,
is that it turns out that in order to get a Hall effect, a phonon exchange is needed
around the intermediate state. Technically, this introduces a crucial phase of
$\pi/2$ (due to a factor i related to an energy conserving δ-function) in the amplitude of going through the intermediate state. A finite phase is needed in that
amplitude in order to get a contribution of the correct (odd) symmetry in Φ,
for the Hall effect. It appears that this crucial phase change along the path is
the quantum analog of the classical velocity, needed to get a Lorentz force,
which brings the Hall effect about.

The above point can be more easily understood by considering a simpler
model for the Hall-type effect, based on a model suggested by Nguyen, Spivak
and Shklovskii (1985) for the magnetoresistance in the hopping conduction
regime. The Hall effect on this model (the simplest version of which is depicted
in Fig. 14 — two tight-binding ordered chains joined by a single tight-binding
plaquette through which a flux Φ is applied — the notation is explained in the
figure) was calculated by Entin-Wohlman, Hartsztein and Imry (1986), the

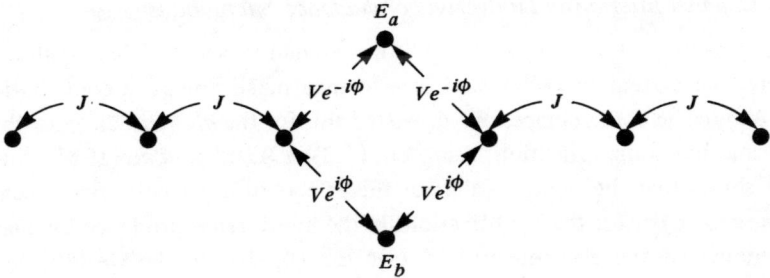

Fig. 14. The simplest version of the Nguyen, Spivak and Shklovskii model, where the sites a and b of the plaquette have random energies E_a and E_b. The tight-binding matrix elements on the ideal leads are J and in the ring, V. A flux is applied to the ring. The Hall-type effect was calculated in this model by Entin-Wohlman *et al.* (1986).

result for the effective R_{yx} being

$$R_{yx} \equiv \frac{V_{ab}}{I} = \frac{\hbar\pi}{e^2} \frac{(\epsilon_1 + \epsilon_2)\sin q \sin\phi}{(\epsilon_1^2 + \epsilon_2^2 + 2\epsilon_1\epsilon_2\cos\phi)} \quad , \tag{5.1}$$

where $\phi = 2\pi\Phi/\Phi_0$, $\epsilon_1 = V^2 J^{-1}(E-E_a)^{-1}$, $\epsilon_2 = V^2 J^{-1}(E-E_b)^{-1}$ and E is the tight-binding energy $E = 2J\cos q$, along the ideal chains. We note the linear dependence on Φ at small Φ, the peridocity in Φ with period Φ_0 and the proportionality to $\sin q$. Since the group velocity along the chains is $v_g = (2J/\hbar)\sin q$, the Hall effect is proportional to v_g.

An interesting and somewhat subtle aspect of the above calculation had to do with the definition of the chemical potentials on the upper and lower sites a and b, whose difference is the voltage V_{ab}. A variant of the "measurement reservoir" concept (Engquist and Anderson, 1981), discussed in Sec. 4.2, had to be introduced for this case. Instead of using the plane wave (here, the Bloch tight-binding states) on the two arms, as in the usual Landauer picture, the exact scattering states were used. Their populations are very simple. The ones coming from the left (right) are full up to an energy $\mu_1(\mu_2)$, the chemical potential of the left (right) particle reservoir. On the other hand, the matrix elements of the coupling of the measurement reservoirs to these scattering states have to be proportional to the magnitude of the latter's wave functions at the measured sites. These two observations enable one to do the calculation of μ_a and μ_b in analogy to the method outlines in Sec. 4.2. As a check, it is straightforward to rederive the Landauer conduction formula using those scattering states. The recent four-terminal calculation of Büttiker (1986b) can also be used to calculate Hall-type resistances.

5.2. *Channel filtering and universal conductance fluctuations*

An argument of Büttiker *et al.* (1985), sketched in Sec. 4.5, shows that in an $N_\perp$ channel system the effects of interference make a relative contribution of $O(1/N_\perp)$ to the conductance. We discussed this for the h/e-periodic contribution in a ring, but some reflection, using Eq. (5.3) of Büttiker *et al.* (1985), immediately shows that the same is valid for the effects of a magnetic field inside the wire (Sec. 4.6) or for the modifications in the interference produced by changing the energy of the electron by $\sim \Delta$ (see Eq. (4.27)). In the weakly localized regime, it was discovered by Altshuler (1985), by Lee and Stone (1985) and by Altshuler and Khmel'nitskii (1985) in important recent papers that these ideas can be clarified and quantified by considering the conductance fluctuations. These may be described by the correlation function, introduced by Lee and Stone:

$$F(\Delta E_F, \Delta H) = \langle g(E_F, H)\, g(E_F + \Delta E_F, H + \Delta H)\rangle - \langle g(E_F, H)\rangle^2 \quad ,$$

$$(5.2)$$

where g is the dimensionless conductance (measured in units of e^2/h) and the angular brackets denote ensemble averaging. Lee and Stone found that this correlation function decays over ranges of E_F and H: $\Delta E_F \sim \Delta$, $\Delta H \sim \Phi_0/A$ (A being here the area of the specimen), as expected from the above considerations. For the magnitude of the conductance fluctuation, both Altshuler and Lee and Stone obtained, in the weakly localized regime ($g \gg 1$) the remarkable result:

$$F(0, 0) = C \quad ,$$

$$(5.3)$$

where C is a numerical constant that depends weakly on the specimen's shape (i.e., $C = .862$ for a square, $C = .729$ for the quasi 1D case, Lee and Stone (1985)), but neither on its size nor on the disorder (hence the term "universal"). These analytical results were corroborated by numerical simulations. There is a growing number of experimental results (Umbach *et al.*, 1984; Jackel, 1983; Blonder, 1984; Skocpol *et al.*, 1985; Licini, Dolan and Bishop, 1985; Licini, Bishop, Kastner and Melngailis, 1985; Kaplan and Hartstein, 1986; Skocpol *et al.*, 1986) (including those on Φ_0-periodic oscillations) which tend to at least semi-quantitatively support these predictions. Several qualitative aspects that were mentioned before, such as the insensitivity of the Φ_0-periodic oscillation in a ring to the field H in the wires, also follow from this formulation.

Evaluating the relative conductance fluctuations from Eq. (5.3) one finds

$$\frac{\Delta g}{g} \sim \frac{1}{N_\perp} \frac{L}{l} \quad ,$$

$$(5.4)$$

recovering the expected $O(1/N_\perp)$ dependence. However, the additional factor $L/l \gg 1$, although in keeping with the lower bound nature of the $\sim 1/N_\perp$ result (Büttiker *et al.*, 1985) is new and unexpected from the considerations presented so far. Actually, the analysis of Büttiker *et al.* (1985) shows that under the assumption that N_{eff} channels contribute each the same order of magnitude to g (we shall call such channels "active channels"), in fact $\Delta g/g \sim 1/N_{\text{eff}}$, so that Eq. (5.4) means that

$$N_{\text{eff}} \sim N_\perp \frac{l}{L} , \tag{5.5}$$

which is valid at this stage only in the weakly localized regime, $L \ll \xi$. This suggests that the number of active channels drops continuously from the full number, $N_\perp$, on microscopic scales to $N_{\text{eff}} \sim 1$ for $L \sim N_\perp l$. The latter length is just the well-known localization length, ξ, as found by Thouless (1977) and discussed in Sec. 3 (we remind the reader (see Eq. (3.10)) that ξ is the length at which $g \sim O(1)$, which is equivalent to $\xi \sim N_\perp l$). Thus $N_{\text{eff}} \sim 1$ is already obtained by stretching Eq. (5.5) to its limit of validity.

The validity of Eq. (5.5) for $L \sim \xi$ is actually almost well known from inedpendent considerations (Azbel, 1982). We now use this to interpret the Altshuler, Lee-Stone result and extend its range of validity. As emphasized by Pichard and Sarma (1981) and Azbel (1982), the electron transmission through the system can be expressed via a multiplicative transfer (or T) matrix. The S matrix needed to calculate g can be computed from the matrix T, where it is more convenient to consider the Hermitean matrix $TT^\dagger$. Oseledec's theorem (1968) guarantees that the eigenvalues of $TT^\dagger$ are Lth powers of a certain set of eigenvalues, out of which only the largest one survives for $L \sim \xi$. g is then essentially given (Azbel, 1982) by the Lth power of that eigenvalue. To connect with the $\exp(-L/\xi)$ decay (Thouless, 1977) of g for $L > \xi$ it must be that this largest eigenvalue is

$$\lambda_{\text{max}} = 1 - \frac{\text{const}}{\xi} \tag{5.6}$$

(all lengths are measured in microscopic units). For $L \gtrsim \xi$ the system behaves one-dimensionally.

On lengths much smaller than ξ, there are more surviving eigenvalues. The process of the elimination of more and more eigenvalues with increasing L is called eigenvector filtering (Pichard and Sarma, 1981). Our point is now that each surviving eigenvector is in effect an active conduction channel. Thus (Imry, 1986), their number determines N_{eff}. The largest surviving eigenvalue λ_L, on scale L is defined by

$$(\lambda_L)^L \sim \frac{1}{e}, \qquad \lambda_L \sim 1 - \frac{\text{const}}{L} . \tag{5.7}$$

Assuming a more or less uniform eigenvalue spacing, we find from Eqs. (5.6) and (5.7)

$$N_{\text{eff}}(L) \sim \xi/L \qquad\qquad\qquad (5.8)$$

which, using the above mentioned value for ξ, is equivalent to Eq. (5.5). The connection between the above assumption and known results on eigenvalue distributions (Wigner, 1951, 1955; Mehta and Dyson, 1963; Mehta, 1967) (the same formalism mentioned in Sec. 2!) and the substantiation of the connection between surviving eigenvectors and active conducting channels were done by Imry (1986). We believe that the above discussion does provide a qualitative understanding of the Altshuler and Lee-Stone results, as well as their extension from $L \ll \xi$ to $L \sim \xi$. For $L \gg \xi$, one is (Azbel, 1983) in the transmission resonance range and some averaging of the fluctuations will occur because the eigenvalues of $TT^\dagger$ become uncorrelated (Imry, 1986; Imry and Sivan, 1986b). We emphasize that the above discussion is valid only when all sample dimensions are shorter than both L_Φ and L_T (the latter is defined in Eq. (4.29)). If the resistance of the sample is measured over times such that the impurities may move, the resistance may fluctuate in time. This is reminiscent of models (Dutta and Horn, 1981) for $1/f$ noise, and may offer a specific model for such noise in small systems. Indeed, very recently Feng, Lee and Stone considered the sensitivity of g to change of a single impurity and considered the ensuing low-frequency noise. Finite temperature effects were considered by Altshuler and Khmel'nitskii (1985), Imry (1986) and Lee *et al.* (1986), and appear to agree with experiment (Licini *et al.*, 1985; Kaplan and Hartstein, 1986; Skocpol *et al.*, 1986).

5.3. *Further problems*

While we concentrated in this paper on the quantum interference phenomena in mesoscopic structures, there exist many further examples of relevant systems and a wealth of interesting phenomena whose exploration promises a stimulating period of research. Still in the electronic realm, the quantum well systems offer many further possibilities, including ballistic transport, the quantized Hall effect and, switching the relevant dimension of the system among the $d = 0, 1, 2$ cases. With small clusters and particles one will be able to see the buildup of the macro-scopic limit, as well as to make contact with the level statistics ideas mentioned in Sec. 2. Small tunnel junctions, both normal and superconducting, can show "macroscopic" quantum tunneling (MQT) (Caldeira and Leggett, 1983) and other extremely interesting phenomena (Ben-Jacob *et al.*, 1983; Ben-Jacob and Gefen, 1985; Likharev and Zorin, 1985; Averin and Likharev, 1985; Widom *et al.*, 1982) due to its small-size. We believe this field is only at its beginning. The electronic properties of large macromolecules, including those that perform biological functions as well as the almost continuous transition to man-made small devices indicate some fundamental questions. Soon, one may understand in detail how an optical transition may cause an electronic excitation to move and trigger the motion of ions or molecules, with various permutations of such processes.

There are also several phenomena which are not necessarily electronic that are important in mesoscopic systems. Many of these are related to the more important role that fluctuations (statistical, classical or quantum) may play in such systems. As an example, we note that the various thermodynamic quantities of a small system weakly coupled to the environment are well defined only *on the average*. However, their instantaneous values fluctuate. When properly defined (Landau and Lifschitz, 1959), intensive quantities such as the temperature can also be considered as fluctuating (although that may be looked at (Landau and Lifschitz, 1959) as fluctuations in the energy or the degree of excitation of the system). Thus, the temperature of a finite system exhibits in equilibrium the following thermodynamic fluctuation

$$\langle \Delta T^2 \rangle = \frac{k_B T^2}{N c_v} \ , \tag{5.9}$$

where N is the number of atoms and c_v the specific heat per atom. Equation (5.9) gives only the magnitude of these fluctuations; their time dependence is determined by several factors including the strength of the coupling to, for example, an isothermal environment. (For very strong coupling the system essentially fluctuates with the much larger environment and for weak enough coupling the system's own fluctuations are slow enough to be relevant and measurable.)

We should like to emphasize that while Eq. (5.9) is not particularly popular among many theoreticians, it can and does easily yield very strong results. For example, the *first* theory of finite-size scaling near second order transitions (see Fisher, 1971) was done by D. J. Bergman and the author (Imry and Bergman, 1971) using Eq. (5.9) and its analogues. The idea has been that since the fluctuations limit how close one can get to the transition in a finite system (the transition as a function of temperature will be rounded by $(T - T_c)_{min}$ $\sim (\langle \Delta T^2 \rangle)^{1/2}$), near this broadened transition the correlation length ξ should reach its maximal possible value, the linear size of the system, and other quantities (e.g., the appropriate susceptibility) will have finite peak values going like *some powers of N*. This can provide a full description of finite size scaling, along with a qualitative understanding how when $N \rightarrow \infty$, the transition sharpens and the finite peaks become mathematical singularities. This picture can also be used to derive (Imry and Bergman, 1971) bulk scaling laws among different critical exponents. For example, when c_v diverges, "hyperscaling" is immediately obtained from Eq. (5.9). When only a higher order derivative of c_v diverges, one has to go to the appropriate higher order fluctuation formula (Landau and Lifschitz, 1959) to establish the scaling. This picture can also be used to derive the rounding of first-order transitions (Imry, 1980b), and to understand how, for example, a thin film crosses over between 2D and 3D behaviors (similar considerations apply also in the electronic transport problem) and to obtain expressions (Imry *et al.*, 1973) that connect critical exponents in different dimensions. While these expressions are not exact, their accuracy is surprising in many cases. Finally, one should mention that in a finite system, the correlation range ξ is anyway limited by the finite size. Thus, even when long-range order does not exist in the "thermodynamic limit", the finite system may be as ordered as if order existed in the bulk (Imry, 1969). This is exactly the same mechanism as the more recent one explaining why thin films made from a metal appear metallic, while "strictly speaking" they are not (Abrahams *et al.*, 1979). The relevant scale here as the smallest of the system's size, L, the length L_Φ, etc.

An interesting and very speculative consequence of Eq. (5.9) emerges at low temperatures (Imry, 1985; Gunther and Ford, 1985). We write

$$\frac{\langle \Delta T \rangle^2}{T^2} = \frac{1}{N(c_v/k_B)} \ . \tag{5.10}$$

Now, consider a small system, very weakly coupled to an isothermal environment. Since c_v vanishes as T approaches zero, the RHS of Eq. (5.10) will exceed unity below a certain temperature T_m. T_m is on the order of $10\,\mathrm{mK}$ for a $(300\,\mathrm{A})^3$ metallic particle, and is much larger than that for an insulating system. Let us now attempt to cool the particle below T_m by reducing the temperature of the environment. The particle's effective temperature, measured by the degree of excitation of its levels, will fluctuate so much that such cooling appears impossible. Thus, *if* ordinary thermodynamic formulas are valid, T_m is the minimum temperature that the particle can in effect be cooled to.

Interesting differences between mesoscopic and macroscopic systems may exist also in metastable and dynamic (e.g., barrier crossing) behavior. On the one hand, for the simple case of homogenous nucleation (Landauer and Swanson, 1961; Langer, 1971), the probability for nucleation is proportional to the volume of the system, which makes homogenous nucleation more probable in a large system. On the other hand, inhomogenous nucleation, for example near the surface, is frequently the dominant mechanism and might be as probable, or stronger, in the small system. A more important case is when the free energy surface itself (i.e., both the structure of the minima and the barriers among them) depends on the size, for example, in a glass or a spin-glass (Sompolinsky *et al.*, 1984) where the barriers may increase with some power of N. A simpler case is that of superparamagnetic particles where, in the presence of anisotropy, the barrier between the "up" and "down" states is $O(N)$. In such cases, the small system may obviously make fluctuations that are practically ruled out in the macroscopic limit. Clearly, the study — experimental and theoretical — of the metastable states and their dynamics is of decisive importance in such systems.

This brings us to the point that various numerical simulations: classical (Binder, 1978, 1984) and quantum (Hirsch and Scalapino, 1984) Monte Carlo, equations of motion (molecular dynamics), master equations, exact computations of the transmission in the transport problem (Stone, 1985) or transfer matrix methods in statistical mechanics (Nightingale, 1976) are possible. They treat finite size systems but can handle interactions, various types of disorder, inhomogeneities, etc. Checking the size-dependence of the various quantities (Imry and Bergman, 1971) is one of the powerful methods to study these systems. Direct comparison with experiments is also possible.

Finally, in addition to the various novel regimes of transport theory, mentioned in the beginning of this chapter, we should like to mention two further, fundamental problems that have to be treated theoretically. The first is the role of fluctuations and noise not only in equilibrium but also, for example, around current-carrying states (Ben-Jacob *et al.*, 1982). The second is the full quantum theory of passage through a potential barrier (where the classical case was solved by Kramers (1940), see also Büttiker, Harris and Landauer (1983), Carmeli and Nitzan (1983)), including the discrete levels in the well(s), the value of the final density of states, when the final state is in the continuum, and/or the coupling to the heat bath or to the other degrees of freedom, which may sometimes be modelled by a set of oscillators. This leads to questions analogous to those discussed in Sec. 3.2, as well as relating to quantum mechanical tunneling, thermal excitation and zero point motion. Some of these questions arise in the MQT (Caldeira and Leggett, 1983) and Polaron (Holstein, 1961) problems; the answers are needed in many physical and chemical systems. Notably, the motion of ions in protein molecules (Frauenfelder and Wolynes, 1985) and through membranes (these being "mesoscopic," at least in some ranges of the parameters) is governed by these considerations. It is interesting that one of the earlier discussions of the important effects of a specific arrangement of units, in contradistinction to ensemble averaging, has been by Azbel (1973) in the context of DNA melting. This difference between specific system properties and those obtained by ensemble (or energy) averaging is a fundamental property of mesoscopic systems, and one which is responsible for many of their interesting and unusual properties.

Acknowledgments

Portions of this research have been supported by grants from the Fund for Basic Research of the Israeli Academy of Sciences and from the U.S.-Israel Binational Science Foundation. The author is extremely grateful to numerous colleagues for constructive and beneficial discussions over the last twenty years. Among these are: Amnon Aharony, Shlomo Alexander, Mark Azbel, Eshel Ben-Jacob, David Bergman, Marc Brodsky, Markus Büttiker, Ora Entin-Wohlman, Alan Fowler, Benjamin Gavish, Yuval Gefen, Leon Gunther, Claudio Hartzstein, Scott Kirkpatrick, Walter Kohn, Rolf Landauer, Patrick Lee, Paul Marcus, Michael Murat, Norman Shiren, Zvi Ovadyahu, Daniel Prober, Douglas Scalapino, Theodore Schultz, Uri Sivan, Douglas Stone, David Thouless, Richard Voss, Sean Washburn and Richard Webb. Those savants who had put too much authoritative emphasis on the generality of the procedure of taking the thermodynamic limit are also thanked, but will remain nameless. Responsibility for errors and misconceptions rests solely with the author. Last, but not least, my deep thanks are due to my late friend and colleague Shang-keng Ma to whom this article is dedicated, for his interest and for many discussions (including one shortly before his death). I am certain that his insights and original ideas on statistical mechanics (including finite systems) will play an important role in the elucidation of the problems mentioned toward the end of this article.

References

Abrahams, E., Anderson, P. W., Licciardello, D. C. and Ramakrishnan, T. V. (1979) *Phys. Rev. Lett.* **42**, 673.

Aharonov, Y. and Bohm, D. (1959) *Phys. Rev.* **115**, 485.

Altshuler, B. L. (1985) *JETP Lett.* **41**, 649.

Altshuler, B. L. and Aronov, A. G. (1985) in *Electron-electron Interactions in Disordered Systems*, Pollak, M. and Efros, A. I., eds. (North-Holland, Amsterdam).

Altshuler, B. L., Aronov, A. G., Khmel'nitskii, D. E. and Larkin, A. I. (1982) in *Quantum Theory of Solids*, Lifschitz, I. M., ed. (Mir Publishers, Moscow) p. 146.

Altshuler, B. L., Aronov, A. G. and Spivak, B. Z. (1981) *JETP Lett.* **33**, 94.

Altshuler, B. L., Aronov, A. G., Spivak, B. Z., Sharvin, D. Yu and Sharvin, Yu. V. (1982) *JETP Lett.* **35**, 588.

Altshuler, B. L. and Khmel'nitskii, D. E. (1985) *JETP Lett.* **42**, 359.

Andereck, B. S. (1984) *J. Phys.* **C17**, 97.

Anderson, P. W. (1981) *Phys. Rev.* **B23**, 4828.

Anderson, P. W., Thouless, D. J., Abrahams, E. and Fisher, D. S. (1980) *Phys. Rev.* **B22**, 3519.

Ando, T., Fowler, A. B. and Stern, F. (1982) *Rev. Mod. Phys.* **59**, 437.

Azbel, M. Ya (1973) *Phys. Rev. Lett.* **31**, 589.

Azbel, M. Ya (1981) *J. Phys.* **C14**, L225.

Azbel, M. Ya (1982) *Phys. Rev.* **B26**, 4735.

Azbel, M. Ya (1983) *Solid State Commun.* **45**, 527.

Azbel, M. Ya and Soven, P. (1983) *Phys. Rev.* **B27**, 831.

Ben-Jacob, E., Bergman, D. J., Imry, Y., Matkowsky, B. J. and Schuss, Z. (1983) *App. Phys. Lett.* **42**, 1045.

Ben-Jacob, E., Bergman, D. J., Matkowksy, B. J. and Schuss, Z. (1982) *Phys. Rev.* **A26**, 2805.

Ben-Jacob, E. and Gefen, Y. (1985) *Phys. Lett.* **108A**, 289.

Benoit, A. D., Washburn, S., Umbach, C. P., Laibowitz, R. B. and Webb, R. A. (1986) submitted for publication.

Bergman, D. J. (1983) private communication; the author is indebted to D. J. Bergman for this suggestion.

Bergmann, G. (1984) *Phys. Reports* **107**, 1.

Binder, K. (1978), ed., *Monte-Carlo Methods in Statistical Physics,* (Springer-Verlag, Berlin).

Binder, K. (1984), ed. *Application of Monte-Carlo Methods in Statistical Physics,* (Springer-Verlag, Berlin).

Binning, G., Rohrer, H., Gerber, Ch. and Weibel, E. (1982) *Phys. Rev. Lett* **49**, 57.

Bishop, D. J., Licini, J. C. and Dolan, G. J. (1985) *App. Phys. Lett.* **46**, 1000.

Bloch, F. (1970) *Phys. Rev.* **B2**, 109.

Blonder, G. (1984) *Bull. Am. Phys. Soc.* **29**, 535.

Brandt, N. B., Bogachek, E. N., Gitsu, D. V., Gogadze, G. A., Kulik, I. O., Nikolaeva, A. A. and Ponomarev, Ya. G. (1976) *JETP Lett.* **24**, 273.

Brandt, N. B., Bogachek, E. N., Gitsu, D. V., Gogadze, G. A., Kulik, I. O., Nikolaeva, A. A. and Ponomarev, Ya. G. (1982) *Sov. J. Low Temp. Phys.* **8**, 358.

Browne, D. A., Carini, J. P., Muttalib, K. A. and Nagel, S. R. (1984) *Phys. Rev.* **B30**, 6798.

Browne, D. A. and Nagel, S. R. (1985) *Phys. Rev.* **B32**, 8424.

Büttiker, M. (1985a) private communication.

Büttiker, M. (1985b) p. 529, Hahlbohm, H. D. and Lubbig, H.

Büttiker, M. (1985c) *Phys. Rev.* **B32**, 1846.

Büttiker, M. (1986a) *Phys. Rev.* **B33**, 3020.

Büttiker, M. (1986b), submitted for publication.

Büttiker, M., Harris, E. P. and Landauer, R. (1983b) *Phys. Rev.* **B28**, 1268.

Büttiker, M. and Imry, Y. (1985) *J. Phys.* **C18**, L467.

Büttiker, M. and Imry, Y. (1986) unpublished results.

Büttiker, M., Imry, Y. and Azbel, M. Ya (1984) *Phys. Rev.* **A30**, 1982.

Büttiker, M., Imry, Y. and Landauer, R. (1983) *Phys. Lett.* **96A**, 365.

Büttiker, M., Imry, Y., Landauer, R. and Pinhas, S. (1985) *Phys. Rev.* **B31**, 6207.

Byers, N. and Yang, C. N. (1961) *Phys. Rev. Lett.* **7**, 46.

Caldeira, A. O. and Leggett, A. J. (1983) *Ann. Phys.* **149**, 374.

Carini, J. P., Muttalib, K. A. and Nagel, S. R. (1984) *Phys. Rev. Lett.* **53**, 102.

Carmeli, B. and Nitzan, A. (1983) *Phys. Rev. Lett.* **51**, 233.

Casimir, H. B. G. (1945) *Rev. Mod. Phys.* **17**, 343.

Castaing, B. and Nozieres, P. (1980) *J. de Physique* (Paris) **41**, 701.

Castaing, B. and Nozieres, P. (1985) private communication. The author is indebted to Drs. Castaing and Nozieres for a discussion on this point.

Chambers, R. G. (1960) *Phys. Rev. Lett.* **5**, 3.

Chandrasekhar, V., Rooks, M. J., Wind, S. and Prober, D. E. (1985) *Phys. Rev. Lett.* **15**, 1610.

Cutler, M. and Mott., N. F. (1969) *Phys. Rev.* **181**, 1336.

Czycholl, G. and Kramer, B. (1979) *Solid State Commun.* **32**, 945.

Datta, S., Melloch, M., Bandyopadhyay, S., Noren, R., Vaziri, M., Miller, M., and Reifenberger, R. (1986a) *Phys. Rev. Lett.* **55**, 2344.

Datta, S., Melloch, M., Bandyopadhyay, S., and Lundstromm, S. (1986b) *Appl. Phys. Lett.* **48**, 487.

Dingle, R. B. (1952) *Proc. Phys. Soc.* **A212**, 47.

Dolan, G. J., Licini, J. C. and Bishop, D. J. (1986) *Phys. Rev. Lett.* **56**, 1493.

Dutta, P. and Horn, P. M. (1981) *Revs. Mod. Phys.* **53**, 497.

Dyson, F. J. (1962) *J. Math. Phys.* **3**, 140, 157, 166.

Economou, E. N. and Soukoulis, C. M. (1981) *Phys. Rev. Lett.* **46**, 618.

Edwards, J. T. and Thouless, D. J. (1972) *J. Phys.* **C5**, 807.

Engquist, H. L. and Anderson, P. W. (1981) *Phys. Rev.* **B24**, 1151.

Entin-Wohlman, O., Hartsztein, K. and Imry, Y. (1986) *Phys. Rev.* **B** in press.

Feng, S., Lu, P. A. and Stone, A. D. (1986) *Phys. Rev. Lett.* **56**, 1960.

Fisher, D. S. and Lee, P. A. (1981) *Phys. Rev.* **B23**, 6851.

Fisher, M. E. (1971) in *Proceedings of the International Summer School Enrico Fermi*, Varenna, Italy, Course 51, M. S. Green, ed. (Academic Press, New York).

Fowler, A. B., U.S. Patent 4550330 (October 1985).

Fowler, A. B., Hartstein, A. and Webb, R. A. (1982) *Phys. Rev. Lett.* **48**, 196.

Frauenfelder, H. and Wolynes, P. G. (1985) *Science* **229**, 337.

Fukuyama, H., (1983) in *Percolation, Localization and Superconductivity*, Goldman, A. M. and Wolf, S. A., eds., Nato Advanced Science Institutes, Series B: Physics 109 (Plenum, New York).

Gefen, Y. (1984) private communication in March.

Gefen, Y., Imry, Y. and Azbel, M. Ya (1984a) *Phys. Rev. Lett.* **52**, 129.

Gefen, Y., Imry, Y. and Azbel, M. Ya (1984b) *Surf. Sci.* **142**, 203.

Gijs, M., van Haesendonck, C. and Bruynseraede, V. (1984) *Phys. Rev. Lett.* **52**, 5069; (1985) *Phys. Rev.* **B30**, 2964.

Gogolin, A. A. and Zimanyi, J. T. (1983) *Solid State Commun.* **46**, 469.

Gordon, J. M. (1984) *Phys. Rev.* **B30**, 6770.

Gor'kov, L. P. and Eliashberg, G. M. (1965) *JETP* **21**, 940.

Gor'kov, L. P., Larkin, A. I. and Khmel'nitskii, D. E. (1979) *JETP Lett.* **30**, 288.

Greenwood, J. (1958) *Proc. Phys. Soc.* London **71**, 585.

Gunther, L. and Ford, L. (1985) unpublished.

Gunther, L. and Imry, Y. (1969) *Solid State Commun.* **7**, 1391.

Hahlbohm, H. D. and Lübbig, H., eds. (1985) *Proceedings of the Third International Conference on Superconducting Quantum Devices*, Berlin, (de Gruyter, Berlin) contains recent references.

Heiblum, M., Nathan, M. I., Thomas, D. E. and Knoedler, C. M. (1985) *Phys. Rev. Lett.* **55**, 2200.

Hikami, S., Larkin, A. I. and Nagaoka, Y. (1980) *Progr. Theor. Phys.* **63**, 707.

Hirsch, J. E. and Scalpaino, D. J. (1984) *Phys. Rev. Lett.* **53**, 2327.

Holstein, T. (1959) *Ann. Phys.* (N.Y.) **8**, 325, 343.

Holstein, T. (1961) *Phys. Rev.* **124**, 1329.

Howard, R. E. and Prober, D. E. (1982) in *VLSI Electronics: Microstructure Science* (Academic, New York) Vol. 5, Chap. 9.

Imry, Y. (1969) *Ann. Phys.* (N.Y.) **51**, 1.

Imry, Y. (1977) *Phys. Rev.* **B15**, 4478.

Imry, Y. (1980a) *Phys. Rev. Lett.* **44**, 469.

Imry, Y. (1980b) *Phys. Rev.* **B21**, 2042.

Imry, Y. (1981) *Phys. Rev.* **B24**, 1107.

Imry, Y. (1983a) *J. Phys.* **C16**, 3501.

Imry, Y. (1983b) in *Percolation, Localization and Superconductivity*, Goldman, A. M. and Wolf, S. A., eds., Nato Advanced Science Institutes, Series B: Physics 109 (Plenum, New York).

Imry, Y. (1985) unpublished.

Imry, Y. (1986) *Europhysics Lett.*, **1**, 249.

Imry, Y. and Shiren, N. (1986) *Phys. Rev.* **B**, in press.

Imry, Y. and Bergman, D. J. (1971) *Phys. Rev.* **A3**, 1416.

Imry, Y., Bergman, D. J., Deutscher, G. and Alexander, S. (1973) *Phys. Rev.* **A7**, 744.

Jackel, L. D. (1983) *Bull. Am. Phys. Soc.* **28**, 401.

Jansen, N. J. M., van Gelder, A. P., Duif, A. M., Wyder, P. and d'Ambrumenil, N. (1983) *Helv. Phys. Acta* **56**, 209.

John, S., Sompolinsky, H. and Stephen, M. J. (1983) *Phys. Rev.* **B27**, 5592.

Kampen van, N. G. (1976) in *Statistical Physics, Proceedings of the IUPAP International Conference*, Pel, L. and Szepflanszy, eds. (North-Holland, Amsterdam).

Kaplan, S. B. and Hartstein, A., (1986) *Phys. Rev. Lett.* **56**, 2403.

Kohn, W. (1964) *Phys. Rev.* **133**, A171.

Kramers, H. A. (1940) *Physica* **7**, 284.

Kubo, R. (1957) *J. Phys. Soc.* Japan **12**, 570.

Kubo, R. (1962) *J. Phys. Soc.* Japan **17**, 975.

Kulik, I. O. (1970) *JETP Lett.* **11**, 275.

Ladan, F. R. and Maurer, C. R. (1983) *C. R. Acad. Sci.* **297**, 227.

Laibowitz, R. (1983) in *Percolation, Localization and Superconductivity*, Goldman, A. M. and Wolf, S. A., eds., Nato Advanced Science Institutes, Series B: Physics, 109 (Plenum, New York).

Landau, L. D. and Lifschitz, E. M. (1959) *Statistical Physics* (Pergamon, London).

Landauer, R. (1957) *IBM J. Res. Dev.* **1**, 223.

Landauer, R. (1966) unpublished IBM proposal.

Landauer, R. (1970) *Phil. Mag.* **21**, 863.

Landauer, R. (1978) in *Electrical Transport and Optical Properties in Disordered Media*, Garland, J. C. and Tanner, D. B., eds. Proceedings of the Ohio State Conference, AIP Conference Proceedings.

Landauer, R. and Büttiker, M. (1985) *Phys. Rev. Lett.* **54**, 2049.

Landauer, R. and Swanson, J. A. (1961) *Phys. Rev.* **121**, 1668.

Langer, J. S. (1971) *Ann. Phys.* **65**, 53.

Langer, J. S. and Neal, T. (1966) *Phys. Rev. Lett.* **16**, 984.

Langreth, D. C. and Abrahams, E., (1981) *Phys. Rev.* **B24**, 2978.

Laughlin, R. B. (1981) *Phys. Rev.* **B23**, 5632.

Lee, P. A. (1984) *Phys. Rev. Lett.* **53**, 2042.

Lee, P. A. and Fisher, D. S. (1981) *Phys. Rev. Lett.* **47**, 882.

Lee, P. A. and Ramakrishnan, T. V. (1985) *Revs. Mod. Phys.* **57**, 287.

Lee, P. A. and Stone, A. D. (1985) *Phys. Rev. Lett.* **55**, 1622.

Lee, P. A. and Stone, A. D. and Fukuyama (1986) submitted for publication.

Lenstra, D. and van Haeringen, W., (1985) *Physica* **128B**, 26, and references therein.

Licini, J. C., Dolan, G. J. and Bishop, D. J. (1985) *Phys. Rev. Lett.* **54**, 1585.

Licini, J. C., Bishop, D. J., Kastner, M. A. and Melngailis, J. (1985) *Phys. Rev. Lett.* **55**, 2987.

Lifschitz, I. M. and Kirpichnikov, V. Ya. (1979) *JETP* **50**, 499.

Likharev, K. K. and Zorin, A. B. (1985) *J. Low Temp. Phys.* **59**, 347; Averin, D. V. and Likharev, K. K. (1986), *J. Low Temp. Phys.* **62**, 345.

Mackinnon, A. and Kramer, B. (1981) *Phys. Rev. Lett.* **47**, 1546.

Mehta, M. L. (1967) *Random Matrices* (Academic Press, New York).

Mehta, M. L. and Dyson, F. J. (1963) *J. Math. Phys.* **4**, 713.

Merzbacher, E. (1961) *Am. J. Phys.* **30**, 237.

Mühlschlegel, B., (1983) in *Percolation, Localization and Superconductivity*, Goldman, A. M. and Wolf, S. A., eds., Nato Advanced Science Institutes, Series B: Physics, 109 (Plenum, New York).

Murat, M., Gefen, Y. and Imry, Y. (1986) *Phys. Rev.* **B**, in press.

Nguyen, V. L., Spivak, B. Z. and Shklovskii, B. I. (1985) *JETP Lett.* **41**, 42.

Nightingale, M. (1976) *Physica* **83A**, 561.

Onsager, L. (1931) *Phys. Rev.* **38**, 2265.

Oseledec, V. I. (1968) *Trans. Moscow Math. Soc.* **19**, 197.

Ovadyahu, Z. and Imry, Y. (1985) *J. Phys.* **C18**, L19.

Pannetier, B., Chaussy, J., Rammal, R. and Gandit, P. (1984) *Phys. Rev. Lett.* **53**, 718; (1985) *Phys. Rev.* **B31**, 3209.

Peierls, R. (1955) *Quantum Theory of Solids* (Clarendon Press, Oxford) p. 29, comment attributed to W. Shockley.

Pepper, M. and Uren, M. J. (1982) *J. Phys.* **C15**, L617.

Pichard, J. L. (1984) Thesis, Paris, Orsay, No. 2858.

Pichard, J. L. and Sarma, G. (1981) *J. Phys.* **C14**, L127.

Prober, D. (1983) in *Percolation, Localization and Superconductivity*, Goldman, A. M. and Wolf, S. A., eds., Nato Advanced Science Institutes, Series B: Physics, 109 (Plenum, New York).

Shapiro, B. (1983) *Phys. Rev. Lett.* **50**, 747.

Sharvin, Yu V. (1965) *Zh. Exp. Teor. Fiz.* **48**, 984; (1965) *Sov. Phys. JETP* **21**, 655.

Sharvin, D. Yu and Sharvin, Yu V. (1981) *JETP Lett.* **34**, 272.

Shtrikman, S. and Thomas, H. (1965) *Solid State Commun.* **3**, 147.

Sivan, U. and Imry, Y. (1986a) *Phys. Rev.* **B33**, 55.

Sivan, U. and Imry, Y. (1986b) submitted for publication.

Skocpol, W. J., Jackel, L. D., Howard, R. E., Mankiewich, P. M. and Tennant, P. M. (1985) to be published.

Skocpol, W. J., Jackel, L. D., Hu, E. L., Howard, R. E. and Felter, L. A. (1982) *Phys. Rev. Lett.* **49**, 951.

Skocpol, W. J., Mankiewich, P. M., Howard, R. E., Jackel, L. D., Tennant, P. M. and Stone, A. D. (1986), *Phys. Rev. Lett.* **56**, 2865.

Sompolinsky, H., Kotliar, G. and Zippelius, A. (1984) *Phys. Rev. Lett.* **52**, 392.

Soukoulis, C. M., Webman, I., Grest, G. S. and Economou, E. N. (1982) *Phys. Rev.* **B26**, 1838.

Stone, A. D. (1985) *Phys. Rev. Lett.* **54**, 2692.

Stone, A. D. and Imry, Y. (1986) *Phys. Rev. Lett.* **56**, 189.

Thouless, D. J. (1977) *Phys. Rev. Lett.* **39**, 1167.

Thouless, D. J. and Kirkpatrick, S. (1981) *J. Phys.* **C14**, 235.

Tonomura, A., Matsuda, T., Suzuki, R., Fukuhara, A., Osakabe, N., Umezaki, H., Endo, J., Shinogawa, K., Sugita, Y. and Fujiwara, H. (1982) *Phys. Rev. Lett.* **48**, 1443.

Umbach, C. P., Van Haesendonck, C., Laibowitz, R. B., Washburn, S. and Webb, R. A. (1986) *Phys. Rev. Lett.* **56**, 386.

Umbach, C. P., Washburn, S., Laibowitz, R. B., and Webb, R. A. (1984) *Phys. Rev.* **B30**, 4048.

Uwaha, M. and Nozieres, P. (1985) *J. de Physique* (Paris) **46**, 109.

Van Vleck, J. H. and Weisskopf, V. F. (1945) *Revs. Mod. Phys.* **17**, 227.

Von Klitzing, K. (1985) private communication, has pointed out to the author that the Onsager relation implies $\sigma(H) = \sigma(-H)$, which implies $G(H) = G(-H)$ only for homogenous systems. Otherwise, the (antisymmetric in H)

Hall part of the conductance tensor might come in and contribute to the effective *G*. The author is indebted to K. von Klitzing for pointing this out to him.

Von Klitzing, K., Dorda, G. and Pepper, M. (1980) *Phys. Rev. Lett.* **45**, 494.

Washburn, S., Umbach, C. P., Laibowitz, R. B. and Webb, R. A., (1985) *Phys. Rev.* **B32**, 4789, and to be published.

Webb, R. A., Washburn, S., Umbach, C. P. and Laibowitz, R. B. (1984) in *Localization, Interaction and Transport Phenomena in Impure Metals*, Bergmann, G., Bruynseraede, Y. and Kramer, B., eds. (Springer-Verlag, Heidelberg) p. 121.

Webb, R. A., Washburn, S., Umbach, C. P. and Laibowitz, R. B., (1985a) to be published in Hahlbohm, H. D. and Lübbig, H. (1985).

Webb, R. A., Washburn, S., Umbach, C. P. and Laibowitz, R. B. (1985b) *Phys. Rev. Lett.* **54**, 2696.

Wheeler, R. G., Choi, K. K., Goel, A., Wisnieff, R. and Prober, D. E. (1982) *Phys. Rev. Lett.* **49**, 1674.

Widom, G., Meglaundis, G., Clarke, T. D., Prance, H. and Prance, R. J. (1982) *J. Phys. Math.* **15**, 156, 3877.

Wigner, E. P. (1951) *Ann. Math.* **53**, 36; (1955) **62**, 548.

Zener, C. (1932) *Proc. Roy. Soc.* **A137**, 636.

MODELS OF PATTERN FORMATION IN FIRST-ORDER PHASE TRANSITIONS

J. S. Langer

Institute for Theoretical Physics
University of California
Santa Barbara, California 93106
USA

The study of simplified theoretical models is an important part of the search for a more general and systematic theory of natural pattern formation. Three qualitatively different kinds of models are considered here, each pertaining to the special class of situations in which patterns are formed during first-order phase transformations. The first of these models is the basic thermodynamic description of a solidifying system controlled by thermal diffusion and interfacial attachment kinetics. The second is a continuum model which is the deterministic limit of "Model C" of Hohenberg, Halperin and Ma. The third is a contour dynamical, boundary-layer model that recently has proven useful for probing the fully nonlinear development of morphological instabilities. Insights concerning the mathematical mechanism for pattern selection in the relatively tractable boundary-layer model appear to be useful clues for understanding the behavior of the more complex and realistic models.

1. The Role of Models in the Search for a Systematic Theory of Pattern Formation

A systematic understanding of the mechanisms by which forms are generated in nature seems well beyond our grasp at present. We see a bewildering variety of patterns emerging even in inanimate phenomena — in solidification, hydrodynamics, chemical reactions, etc. — and we are only just beginning to comprehend the enormity of the task of understanding how patterns are formed in living systems. Eventually, we hope to know how to sort out these processes according to underlying mechanisms, to understand which phenomena are related to one another, to distinguish intrinsically complex situations from those which are fundamentally simple. We are not yet near the point where we can do such things, but it is always useful to think about what we need to learn in order to get there.

The most obvious classification of processes, one which has begun to be studied seriously in recent years, distinguishes physical situations which actually lead to regular, reproducible patterns from those which do not. For example, small slightly supercritical Rayleigh-Bénard convection cells exhibit sharp wavelength selection induced by boundary conditions (Whitehead, 1975). Selection of dendritic shapes and growth rates also seems to be sharp, but is independent of boundaries (Glicksman *et al.*, 1976; Huang and Glicksman, 1981; Langer, 1980). In larger Rayleigh-Bénard systems, wavelength selection is history dependent or nonexistent (Heutmaker *et al.*, 1985; Ahlers *et al.*, 1985). A growing number of theoretical models and experimental situations is currently being studied from this point of view, and I am optimistic that some specific questions will be answered in the next few years. Specific answers should lead to more general conclusions.

Another kind of classification that I think might be achievable , probably on a longer time scale, would involve the amount of information required to specify a pattern-forming event. For example, complicated dendritic structures are formed during solidification of simple substances in which both the intrinsic properties of the material such as thermal conductivity and surface tension as well as externally controlled parameters such as the temperature of the melt are held constant throughout the process. Natural snowflakes, on the other hand, owe some of their diversity to slight changes in the temperature and humidity of their environments during the history of their formation (Hobbs, 1974). In biological processes, the pattern-forming media themselves apparently change their properties in response to complex sequences of external stimuli. Might it not be possible to learn how to recognize some external signatures of these different levels of intrinsic complexity? Equivalently, might it not be possible to deduce from the external features of a pattern forming process what are the minimum ingredients of an underlying physical model of the phenomenon?

Another example of the kind of classification that I think would be useful has to do with the precision with which pattern formation is controlled. Dendrites, convection patterns, and the like are stable against thermal fluctuations, but are never more precise in their features than is consistent with the noise in their environments. Some biological processes, on the other hand, appear to do better[1]. I think that we are on the verge of understanding how some inanimate processes may be acutely sensitive to small systematic changes in growth conditions without actually amplifying environmental noise; but the mechanisms by which living systems approach quantum limits of precision — if indeed they do so — constitute scientific *terra incognita*. As in the case of complexity, systematic characterization of these various levels of precision is equivalent to specifying minimum ingredients of adequate physical models.

If we are far from possessing a systematic theory of morphogenesis, the reason, I think, is that we still do not understand very well the properties of even the simplest models of pattern formation. Accordingly, my intention in this essay is to write about one special class of models, those that are based on first-order phase transitions, and to encourage the reader to consider with me how the information that we are obtaining about these models may ultimately fit into a bigger picture. As I shall argue later, association with a true thermodynamic phase transition is not likely to be a deep, distinguishing feature of pattern-forming systems in the sense outlined above. The choice of this class of models simply serves to narrow the discussion to manageable dimensions, and also lets me touch upon some work of Shang-Keng Ma (Halperin *et al.*, 1974; Hohenberg and Halperin, 1977), to whose memory this essay is dedicated.

My plan is to start by describing what I shall call the basic model of solidification. In fact, this is a fairly general model of what happens when two phases of a system are in contact with each other at a well defined boundary, and when the departure from equilibrium between these two phases causes that boundary to move. To emphasize the generality of this physical situation, I shall then move one step backwards toward a more fundamental description by showing how this model is related to the Ginzburg-Landau or Cahn-Hilliard models of phase transitions that have become so familiar to physicists in the last three decades. Finally, I shall move forward in time to describe a drastically truncated

---

[1] The biological literature is so huge and foreign to this author that it does not seem possible to insert an exactly appropriate reference here. A classic text is that of D' Arcy W. Thompson (1944). For a more recent review, see Belintsev (1983a, 1983b). The problem of precision in biological systems has been addressed by Bialek (1983).

"boundary-layer" model (Ben-Jacob et al., 1983, 1984b), that I and my colleagues have proposed recently for the purpose of probing the fully nonlinear behavior of solidifying systems. This newer model has led to some interesting and surprising results which, if confirmed in the case of more realistic systems, may provide useful clues about a more general theory.

2. The Basic Model of Solidification Theory

In the conventional thermodynamic model of the solidification of a pure sub-stance from its melt, the fundamental rate-controlling mechanism is the diffusion of latent heat away from the interface (Chalmers, 1964; Woodruff, 1973). The dimensionless thermal diffusion field is conveniently chosen to be

$$u = \frac{T - T_M}{(L/c)} \; , \tag{2.1}$$

where T_M is the melting temperature and the ratio of the latent heat L to the specific heat c is an appropriate unit of undercooling. The field u satisfies the diffusion equation

$$\frac{\partial u}{\partial t} = D \nabla^2 u \; , \tag{2.2}$$

where D is the diffusion constant which can, if we choose, be taken to have different values in the liquid and solid phases.

The crucial ingredients of the model are the boundary conditions imposed on u at the solidification front. First, there is heat conservation:

$$v_n = - [D \hat{n} \cdot \nabla u] \; , \tag{2.3}$$

where $\hat{n}$ is the unit normal directed outward from the solid, v_n is the normal growth velocity, and the square brackets denote the discontinuity in the heat flux across the boundary. The physically more interesting condition is a state-ment of thermodynamic equilibrium — or small departure therefrom — which determines the temperature u_s at the two-phase interface:

$$u_s = - d_0 \kappa - \beta(v_n) \; . \tag{2.4}$$

The right-hand side of (2.4) describes deviations from the state of thermo-dynamic equilibrium between two bulk phases in contact at a flat interface. The first term is the Gibbs-Thomson correction for a curved surface; κ is the sum of the principle curvatures, and $d_0 = \gamma c T_M / L^2$ is a capillary length, ordinarily of order Ångstroms, which is proportional to the surface tension γ.

The second term in (2.4) accounts for departures from local equilibrium associated with motion of the interface. To see this more generally, rewrite the equation in the form

$$v_n = \beta^{-1}(\delta u_s) , \qquad (2.5)$$

where $\delta u_s \equiv -d_0 \kappa - u_s$ is the effective undercooling at the interface and $\beta^{-1}(\delta u_s)$ is the functional inverse of $\beta(v_n)$. The special choice $\beta^{-1} = \delta u_s/\beta_0$ implies linear response with a kinetic coefficient β_0^{-1}. A linear law of this kind might be accurate for a molecularly rough interface, in which case $\beta_0 = 0$ would describe the limit of pure diffusion control. For facetted interfaces, on the other hand, β^{-1} may be a highly nonlinear function of δu_s. In the latter case it is clear that β^{-1} must depend strongly on the orientation of the interface relative to the axes of symmetry of the growing crystal. Even in the linear, nonfacetted situation, both the inverse kinetic coefficient β_0 and the capillary length d_0 may be orientation-dependent and such effects of crystalline anisotropy seem to be extremely important in pattern selection.

The above equations, supplemented by initial data and boundary conditions on u far from the solidification front, constitute a complete statement of a physically realistic moving-boundary problem. In principle, these equations could be programmed into a computer and, with proper choices of parameters, the computer would predict the snowflake-like patterns seen experimentally when pure substances are solidified from undercooled melts[2]. The model does omit some physical effects such as hydrodynamics of the fluid phase and any elastic, plastic, or dynamic properties of the solid; but I think it highly unlikely that any of these effects will be of central importance for dendritic pattern formation. In practice, however, this model has been too difficult for extensive numerical analysis except in special cases which are, for our purposes, too simple to be interesting. The advent of parallel-processing computers, adaptive-grid algorithms, and the like should improve this situation in the near future, and the ability to make accurate simulations of the basic solidification model with various geometries and choices of parameters ultimately should prove very useful in developing and testing basic theory. I want to emphasize, however, that the role of numerical simulation at best is not essentially different from that of carefully controlled experimentation, especially in situations like this where the underlying physical model seems well understood. Experiments, both real and numerical,

---

[2] So far as I know the closest that anyone has come to seeing dendritic behavior in a numerical solution of the basic solidification model is the work of Oldfield (1973). Oldfield's methods, although subject to some uncertainties, seem to have been about ten years ahead of their time. For a more recent attempt, see Smith (1981).

can test existing theoretical ideas and inspire new ones, but cannot substitute for fundamental theory.

3. The Phase-Field Model

The phase-field model, except for its neglect of stochastic forces, is identical to what Halperin, Hohenberg, and Ma (1974) have called "model C" in their studies of nonequilibrium critical phenomena; that is, it is a phenomenological model of the class of phase transitions which can be described by a nonconserved scalar order parameter coupled to a conserved noncritical thermal field.

There are two special reasons for discussing model C at this point in this essay. First, the classification of critical-point models that has emerged from a fundamental understanding of their distinguishing dynamical characteristics is a good example of the kind of systematization that I should like to see for models of pattern formation. I do not mean to imply that there is any particular similarity between the two kinds of problems, or that the problem of pattern formation as a whole might be amenable to such a complete systematization. However, in the next several paragraphs, we shall observe that the basic solidification model is a limiting case of model C; and this fact may help us to recognize similarities within this class of pattern-forming processes. For example, the role of the rate-limiting thermal field could be played by diffusing impurities, or diffusion of the primary substance through an inert secondary medium, without changing the mathematical statement of the model. On the other hand, this model of a true phase transition appears to be subtly different from, say, diffusion-limited aggregation (Witten and Sander, 1983).

The second reason for current interest in model C is that it may provide the most natural algorithm for numerical solutions of the basic solidification model. This is the context in which the term "phase-field model" has been introduced by Fix (1983)[3]. A serious difficulty with the basic model is that it requires the computer to track a moving boundary, a procedure which is frought with numerical hazards. The idea which Fix has pursued is to replace the dynamics of the boundary by an equation of motion for a phase-field — in other words, an order parameter — which changes from one value to another quickly but smoothly at the two-phase interface. The obvious way of inventing such an equation is to go back to the physics, that is, to use model C. The new partial

---

[3] Fix and Lin (unpublished) have demonstrated the feasibility of their technique for studying the onset of instabilities but have not yet used it to generate more complex solidification patterns. A similar idea has been developed independently by Collins and Levine (1985), and also by Deutsch (1985).

differential equation for the phase field is necessarily much stiffer than the diffusion equation for the thermal field; but stiff equations may turn out to be easier to control numerically than moving boundaries. The results are not known yet.

My purpose in the next few paragraphs is to provide a simple, intuitive description of the phase-field model without any attempt at rigor or completeness. Most of the ideas probably would have seemed familiar to Gibbs or van der Waals, but I think that they are worth repeating in a modern context.

Let $\phi(\mathbf{x}, t)$ be the phase field, a function of position $\mathbf{x}$ and time t. Its equation of motion can be chosen to be

$$\frac{\partial \phi}{\partial t} = -\Gamma \frac{\delta F}{\delta \phi} \ , \tag{3.1}$$

where Γ is a kinetic coefficient and F is a free energy with the standard form:

$$F\{\phi\} = \int d\mathbf{x} \left[\frac{K}{2} (\nabla \phi)^2 + f(\phi) - \alpha u \phi \right] . \tag{3.2}$$

Here, $f(\phi)$ is a free-energy density which we can choose to have minima with $f = 0$ at $\phi = \pm \Delta \phi$, so that the latter quantities are the equilibrium values of ϕ in the coexisting "liquid" $(+)$ or "solid" $(-)$ phases. The gradient-energy coefficient K determines the shape in equilibrium of the interfacial profile $\phi_s(x)$, which satisfies (3.1) with $d\phi_s/dt = 0$ and $u = 0$:

$$K \frac{d^2 \phi_s}{dx^2} = \frac{df}{d\phi_s} \ , \tag{3.3}$$

and has associated with it the interfacial energy (Cahn and Hilliard, 1958)

$$\gamma = K \int_{-\infty}^{\infty} \left(\frac{d\phi_s}{dx} \right)^2 dx \ . \tag{3.4}$$

The coupling coefficient α can be determined from considerations of thermodynamic equilibrium to be $L^2/T_M c \Delta \phi$.

The phase-field model reduces to the basic solidification model in the limit where the thermal field u varies much more slowly than the phase-field ϕ in both space and time. An easy — and easily generalizable — way to see how this works is to consider a growing d-dimensional sphere in a region of constant

$u < 0$, and to look for an interfacial profile of the form $\phi \equiv \phi_s(r - R)$, where r is the radial coordinate, R is the slowly time dependent radius of the sphere, and ϕ_s is almost but not quite the same function that was defined in (3.3). Specifically, for r near R, ϕ_s now is a solution of

$$K \frac{d^2 \phi_s}{dr^2} \simeq \frac{d}{d\phi_s}(f - \alpha u \phi_s) - \eta \frac{d\phi_s}{dr} ,$$ (3.5)

where

$$\eta = \frac{(d-1)K}{R} + \frac{v}{\Gamma} ,$$ (3.6)

and $v = dR/dt$. When $|u|$ is sufficiently small, Eq. (3.5) can be solved by eye by the usual mechanical analogy; r becomes time and ϕ_s becomes the displacement of an object of mass K moving in the potential $-f + \alpha u \phi_s$ with damping constant η. The particle stays for a time R at the "solid" peak near $\phi = -\Delta\phi/2$ and then drops abruptly to the "liquid" peak near $+\Delta\phi/2$, losing an energy $\alpha |u| \Delta\phi$ in the process. The situation is illustrated in Fig. 1. The requirement that the friction constant η be just right to achieve this energy loss is:

$$-\alpha u (\Delta\phi) \simeq \eta \int \left(\frac{d\phi_s}{dr}\right)^2 dr \simeq \frac{\eta \gamma}{K} .$$ (3.7)

In writing the final expression on the right-hand side of (3.7), I have used (3.4) and have ignored the small u-dependence of γ. This result is, in effect, a condition for the solvability of (3.5). It is conveniently rewritten in the form:

$$v = -\frac{(d-1)\Gamma K}{R} - \frac{\Gamma K \alpha u \Delta\phi}{\gamma} .$$ (3.8)

Equation (3.8) is the same as (2.4) or (2.5) in the case of linear response if we identify $\beta_0^{-1} = \Gamma K L^2 / \gamma C T_M$, set $u = u_s$, and generalize $(d-1)/R$ to be the curvature κ. Note that we have also recovered the Gibbs-Thomson condition by this procedure.

It remains to look at variations of the thermal field u. Let σ be the entropy density. Up to a ϕ-independent constant, σ is given by

$$\sigma = -\frac{\partial f}{\partial \tau} = \frac{\alpha C}{L}\phi = \frac{L\phi}{T_M \Delta\phi} ,$$ (3.9)

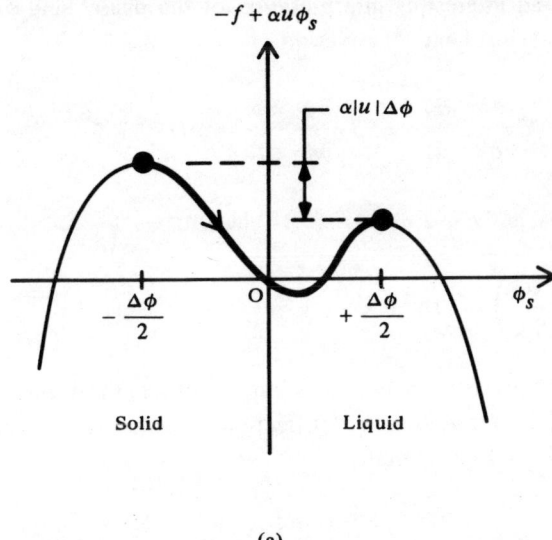

(a)

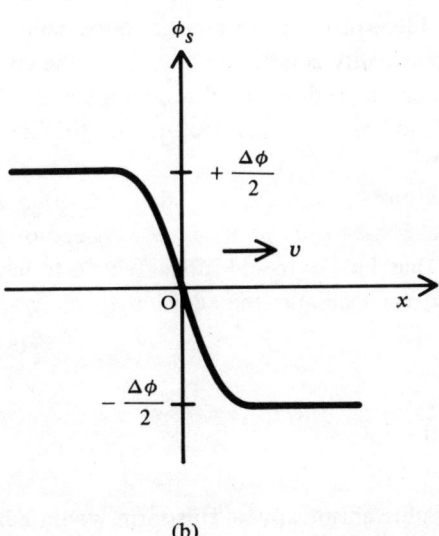

(b)

Fig. 1 (a) Schematic effective potential for the mechanical analogy used in the derivation of the equation of motion (3.8) for an interface in the phase-field model.
(b) Equilibrium interfacial profile $\phi_s(x)$.

which provides an interesting interpretation of the phase-field ϕ. The equation of motion for σ is just heat conservation:

$$T\frac{d\sigma}{dt} = T\frac{\partial\sigma}{\partial T}\frac{dT}{dt} + T\frac{\partial\sigma}{\partial\phi}\frac{d\phi}{dt} = cD\nabla^2 T . \tag{3.10}$$

With (2.1), (3.9), and $c = T\partial\sigma/\partial T$, (3.10) becomes

$$\frac{\partial}{\partial t}\left(u + \frac{\phi}{\Delta\phi}\right) = D\nabla^2 u . \tag{3.11}$$

Away from the interface, ϕ is a constant, and (3.11) reduces to the diffusion equation (2.2). Suppose that a flat section of interface is moving in the x direction with velocity v, so that

$$\phi(x,t) \simeq \phi_s(x - vt) ; \qquad \frac{\partial\phi}{\partial t} \simeq -v\frac{\partial\phi_s}{dx} . \tag{3.12}$$

Remembering that u must be continuous and slowly varying on the scale of the thickness of the interface, we can use (3.12) to integrate (3.11) from a point just on the solid side, say $x \leq 0$, to a neighboring point in the liquid at $x \geq 0$. The result is the continuity condition (2.3). Thus the combination of equations (3.1) and (3.11) does reproduce, in the limit of a sharp interface and slowly varying thermal field, most of the features of the basic solidification model described in Sec. 2.

One crucial feature of solidification that has been missing so far in this description of the phase-field model is any effect of crystalline anisotropy. The simplest way that I know to add such effects is to include higher derivatives in the free energy; for example, the addition to the gradient energy of a term proportional to

$$\sum_{i=1}^{d}\left(\frac{\partial^2\phi}{\partial x_i^2}\right)^2 ,$$

would produce a cubic anisotropy[4]. This term would add fourth derivatives to equations (3.3) or (3.5) but apparently does not change the above results in any essential way other than causing the surface tension to become anisotropic.

---

[4] Deutsch (1985) has demonstrated that a four fold anisotropy can stabilize the direction of growth of an embryonic dendritic protrusion.

The effect is the same as that of adding next-neighbor interactions along the symmetry axes in an Ising-like discretization of this model. One unnatural feature of this procedure is that the anisotropy pertains equally to the "liquid" as well as "solid" phases although it has a strong effect only at the interface. If that turned out to be unacceptable for some purposes, one could make the new term vanish in the "liquid" phase by multiplying it by an appropriate function of ϕ.

Because the phase-field model is a purely thermodynamic description of a first-order phase transition, one might expect that there exists a thermodynamic free energy functional which is a non-increasing function of time and which, therefore, serves as a Lyapunov functional for the coupled system of equations (3.1) and (3.11). Indeed, the appropriate free energy is

$$\mathcal{F} = \int dx \left\{ \frac{K}{2} (\nabla \phi)^2 + f(\phi) + \tfrac{1}{2}\, \alpha \Delta \phi u^2 \right\} . \qquad (3.13)$$

To see this, let the independent fields be ϕ and $U = u + \phi/\Delta\phi$, in terms of which the equations of motion (3.1) and (3.11) can be written

$$\frac{\partial \phi}{\partial t} = -\Gamma \frac{\delta \mathcal{F}}{\delta \phi} , \qquad (3.14)$$

$$\frac{\partial U}{\partial t} = \frac{D}{\alpha \Delta \phi} \nabla^2 \frac{\delta \mathcal{F}}{\delta U} . \qquad (3.15)$$

Then, so long as there are no fluxes across boundaries at infinity,

$$\frac{d\mathcal{F}}{dt} = -\int dx \left\{ \Gamma \left(\frac{\delta \mathcal{F}}{\delta \phi} \right)^2 + \frac{D}{\alpha \Delta \phi} \left(\nabla \frac{\delta \mathcal{F}}{\delta U} \right)^2 \right\} \leqslant 0 ; \qquad (3.16)$$

thus $\mathcal{F}$ is a Lyapunov function.

So far as I know, the variational principle (3.16) is of no special help in solving pattern-selection problems; nor do I know of any variational formulation that is more useful. In systems undergoing phase transformations, patterns are either transient phenomena, that is, nonequilibrium shapes like snowflakes which are *en route* to simpler equilibrium structures, or else they occur in systems which are being driven away from equilibrium by constraints — fluxes at infinity — which invalidate (3.16). An example of the latter situation is directional solidification (Langer, 1980) where one observes, say, cellular solidification fronts in a moving frame of reference in which heat is being added at one end and extracted at the other. The resulting problem is very similar to the Rayleigh-Bénard problem in hydrodynamics (Whitehead, 1975) for which no Lyapunov function seems to exist (except, perhaps, in the limit of weak flow near the threshold of instability). The lesson to be learned from all of this is that the question of whether a pattern-forming system is undergoing a first-order phase transition is *not* likely to be of primary importance in a systematic theory of morphogenesis. The focus of attention must not be on the existence of a free-energy or the like, but rather on the dynamical properties of the equations of motion themselves.

4. The Boundary-layer Model

The feature of the basic solidification model that makes it so difficult mathematically is its spatial and temporal nonlocality. The latent heat released at one point on the solidification front determines the motion of distant points at later times. As a result, a complete mathematical description must retain either the full diffusion field $u(\mathbf{x}, t)$ or, equivalently, the entire history of configurations of the front. Thus, the solidification model is intrinsically different from, for example, models of viscous fingering in porous media or in Hele-Shaw cells where the diffusion equation (2.2) for the thermal field is replaced by the Laplace equation for the pressure (Hele-Shaw, 1898; Saffman and Taylor, 1958). The latter replacement exaggerates spatial nonlocality but eliminates memory effects; the motion of the interface is determined only by its current configuration, and numerical analysis is feasible (Tryggvason and Aref, 1983). If we try omitting $\partial u/\partial t$ in (2.2) to study solidification, however, we lose Ivantsov's steady-state "needle-crystal" solutions (Ivanstov, 1947; Horvay and Cahn, 1961) and, therefore, we probably lose the possibility of studying persistent, i.e., non-transient, dendritic behavior[5].

The boundary-layer model (Ben-Jacob *et al.*, 1983; 1984b) has been invented in an attempt to include some of the nonlocality associated with the thermal diffusion field in an otherwise purely contour-dynamical description of pattern formation. The range of the diffusion field in front of a moving interface is generally of order $l \sim D/v$. If this range is much smaller than the radius of curvature of the interface, $\kappa l \ll 1$, then diffusion is effectively confined to a narrow region which we call the boundary-layer. Our basic idea is to replace the dynamics of the full diffusion field by an equation of motion for the local heat content of the boundary-layer as a function of position on the interface.

---

[5] With the addition of anisotropy, however, these systems have been shown to produce remarkable snowflake-like patterns. See Ben-Jacob *et al.* (1986).

This major simplification (oversimplification?) turns out to have great mathematical benefits. In particular, lateral diffusion within the boundary layer mimics the retarded nonlocal interactions between different points on the solidification front, and these interactions can be described mathematically in a relatively tractable manner.

It must be emphasized that what is being described here is a model, not a systematic approximation. There do exist physical situations in which the condition $\kappa l \ll 1$ might be valid, for example, dendritic tips at large undercooling; although this is not the regime in which accurate experiments have yet been performed. Even in this regime, however, the natural instabilities of the solidification front lead to situations in which curvature is large or in which points on the front that are well separated from each other in arc length — for example, points on different side branches of a dendrite — approach each other in real space. Either of the latter situations imply a breakdown of the boundary-layer assumption. My preference is to think of the boundary-like picture as a model in its own right, guided by the physics of solidification but not expected to produce quantitative experimental predictions, its main purpose being to provide new insights into mathematical mechanisms of pattern selection.

Given the special purposes of the boundary-layer model, it has been reasonable to confine detailed analyses to the two-dimensional situation in which the solidification front is a simply connected curve, i.e., a "string". We can represent this front by specifying its curvature κ as a function of arc length s and time t. If θ is the angle between the normal to the front and a fixed direction (perhaps the orientation of some axis of symmetry of the growing crystal), then the definition of curvature,

$$\kappa(s,t) = \frac{\partial \theta}{\partial s} , \tag{4.1}$$

is a differential equation whose solution determines the curve. The equation of motion for κ is

$$\left(\frac{\partial \kappa}{\partial t} \right)_n = -\left(\frac{\partial^2}{\partial s^2} + \kappa^2 \right) v_n , \tag{4.2}$$

which must be supplemented by the metric condition

$$\left(\frac{ds}{dt} \right)_n = \int_0^s \kappa \, v_n \, ds' . \tag{4.3}$$

Here, the subscript n denotes differentiation along the outward normal and v_n is the same normal velocity defined previously. Equations (4.2) and (4.3) are purely geometric statements; all physical content is contained in the quantity v_n.

The essence of the boundary-layer model is the assumption that $v_n(s,t)$ must depend only on local properties of the system at the position s and time t. The simplest assumption is that v_n is a function only of the local geometry; for example, v_n could have the form

$$v_n = (1 + \epsilon \cos m\theta)\left(\kappa + A\kappa^2 - B\kappa^3 + \gamma \frac{\partial^2 \kappa}{\partial s^2} \right) , \qquad (4.4)$$

where A, B, γ and ϵ are constants and m is an integer, say $m = 4$ or 6, which determines the crystalline symmetry. This is the "geometrical model" of Brower *et al.* (1983; 1984), which has proven to be immensely useful because it is simple enough to be analyzed in great detail and turns out to have interesting — if not quite dendritic — pattern forming properties.

The geometrical model (4.4) is missing the memory effects that I have advertized as being so important. For example, a flat interface ($\kappa = 0$) never moves. In real life, the velocity of such an interface can be finite, but must decrease like $t^{-1/2}$ because of the gradual build-up of latent heat in the fluid ahead of it. A related feature is that, although the geometrical model does have steady-state needle-crystal solutions when the "surface tension" γ vanishes, these needles have stationary straight sides instead of the parabolic shape characteristic of models with diffusion control. The principal success of the geometrical model is that the growth rates of its needle crystals at finite γ are selected *via* a solvability mechanism which can be studied in great detail (Kessler *et al.*, 1984). Disappointingly, these needle crystals do not undergo persistent side branching oscillations except at one special value of the anisotropy strength ϵ for which, as it turns out, the tip of the needle is just marginally unstable. The geometrical model is not, strictly speaking, a direct representation of a first-order phase transition; therefore I shall say no more about it except to emphasize again its importance as a mathematical example.

In the boundary-layer model proposed by Ben-Jacob *et al.* (1983; 1984b), the velocity v_n is determined by l, the thickness of the layer. We suppose that the thermal field u in front of the interface looks something like

$$u(x) \sim -\Delta + (\Delta + u_s)e^{-x/l} , \qquad (4.5)$$

where x is the distance into the fluid measured along the unit normal $\hat{n}$, $\Delta = (T_M - T_\infty)/(L/c)$ is the undercooling at infinity in units L/c, and u_s is the value of u at the interface given by (2.4) or, equivalently, (3.8). The continuity condition (2.3) suggests that we use the ansatz (4.5) to write

$$v_n = \frac{D(\Delta + u_s)}{l} \ . \tag{4.6}$$

It remains to devise an equation of motion for l. This is best done in terms of the quantity

$$h(s, t) = \int_0^\infty dx \, (\Delta + u) \simeq (\Delta + u_s) l \ , \tag{4.7}$$

which we interpret as the heat content per unit length of the interface. Our proposed equation of motion for h is a phenomenological statement of heat balance within the boundary layer:

$$\left(\frac{\partial h}{\partial t} \right)_n = v_n (1 - \Delta - u_s) - v_n \kappa h + D \frac{\partial}{\partial s} \left(l \frac{\partial u_s}{\partial s} \right) . \tag{4.8}$$

Equation (4.8) is the crux of the boundary-layer model, and each of the three terms on the right-hand side deserves some comment. The first term is the rate at which latent heat is being added to the boundary layer. It accounts for the fact that a fraction $\Delta + u_s$ of this heat must be used to warm the solid from $u(\infty) = -\Delta$ to $u = u_s$. The second term is geometrical in origin and contains the mechanism for the diffusive instability that is responsible for pattern formation in these systems (Langer, 1980). It expresses the fact that a piece of interface with positive curvature increases in length as it grows outward, thus decreasing the amount of heat per unit length, h. Because the corresponding decrease in l implies an increase in v_n, this term causes an outward bulge to accelerate. Conversely, the heat h accumulates in an inward pointing groove, causing it to fall behind the neighboring parts of the interface. The last term on the right-hand side of (4.8) describes lateral diffusion, and it is *via* this effect that capillary forces control the diffusive instability just described. Remember that, apart from the kinetic correction, u_s is equal to $-d_0 \kappa$, so that this third term couples the thermal field to the curvature with a strength proportional to the surface tension. More precisely, a forward bulge, $\kappa > 0$, must be cooler than its surroundings by an amount $d_0 \kappa$, and therefore heat must flow toward it, retarding its growth. The specific form chosen for the lateral diffusion term depends strongly on the presumed locality of the diffusion field, and it is probably at this point that the boundary-layer model is least realistic.

The dynamical system defined by Eqs. (4.6) and (4.8), supplemented by the thermodynamic relation (2.4), has many features in common with the basic solidification model of Sec. 2. Generally, the two models are in good agreement when the boundary-layer condition, $\kappa l \ll 1$, is satisfied. Flat interfaces move with the appropriate $t^{1/2}$ law and needle-crystals in the Ivantsov limit, $d_0 = 0$, are parabolas. The dynamics of the diffusive instability, in the long-wavelength limit, is the same in both models. There are also important discrepancies, mostly associated with failure of $\kappa l \ll 1$. For example, stable, short-wavelength deformations of a flat interface relax much too slowly because there is no mechanism in this model to account for the fact that such deformations perturb the diffusion field only out to distances less than l. Another awkward feature of the boundary-layer model is that it does not, in its present form, include diffusion in the solid. As a result, heat accumulating in grooves cannot escape, and the growing instabilities tend toward sharp cusps which may be physically realistic but have been major impediments to numerical analysis.

Although analytic and numerical work on the boundary-layer model has not yet been carried as far as that on the geometrical model (Kessler *et al.*, 1984), a number of very interesting results have been obtained. I want to devote the last few paragraphs of this essay to a qualitative description of some of those results because I think they illustrate the role of model building in the search for more general theories of pattern formation.

One surprising result that emerged early in our numerical studies of the boundary-layer model was the crucial role played by crystalline anisotropy. The existence of Ivantsov's needle-crystal solutions in the absence of anisotropy had led to the expectation that its role, while important as a symmetry-selecting perturbation, was somehow secondary to the dynamical interaction between diffusive instabilities and capillary forces. Our first numerical simulations made it clear that such was not the case.

Figure 2 shows a sequence of solidification fronts computed with no crystalline anisotropy; that is, $\beta = 0$ and d_0 is independent of θ in (2.4). The initial shape is a circle with a small six-fold anisotropy, and six-fold symmetry is enforced at all later stages. Note that we have been able to follow the morphological instability far into the nonlinear regime where the outward bulges flatten and the heat accumulates in deep grooves. This behavior is generally characteristic of unstable solidification fronts observed in experiments but, so far as I know, this is the first time that it has been reproduced theoretically. Also apparent in the figure is that this process does not produce a snowflake. As verified by further computation, the next grooving instability occurs at the flattened centers of the outward bulges, producing a pair of fingers which, themselves, will later flatten and split. If continued indefinitely, this system

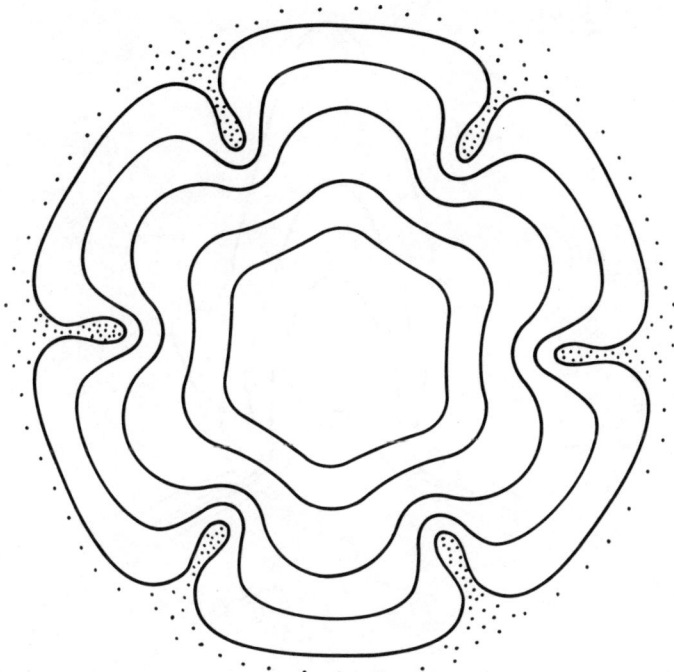

Fig. 2 Sequence of solidification fronts generated by the boundary-layer model with isotropic growth kinetics. The initial state contains a small six-fold anisotropy. The dots indicate schematically the accumulation of heat in the grooves.

would produce an increasingly complex, possibly fractal, structure which might be of some interest in biology but probably not in crystal growth. Moreover, the symmetry of this structure is almost certainly unstable. Although we have not pursued this line of investigation in any detail, I am fairly sure that a small symmetry breaking perturbation would cause this system to produce an intrinsically chaotic picture more reminiscent of seaweed than any regular pattern.

Figure 3 shows what happens when anisotropy is added to the equations of motion. Here we have started with an unperturbed circle (the dashed curve in the figure, one sixth of which is shown); we have let d_0 be constant again, but have chosen a linear kinetic coefficient β_0 proportional to $1 - \cos 6\theta$. The circle first grows slowly almost into a hexagon, then becomes unstable at its corners and emits dendritic structures which propagate rapidly outward along the growth directions favored by $\beta_0(\theta)$. The dynamical effect of the anisotropy is to cause these dendritic tips to lock almost immediately into a state with sharply defined velocity and tip radius, both of the latter parameters being the same as those which are selected when the system is set into motion from different

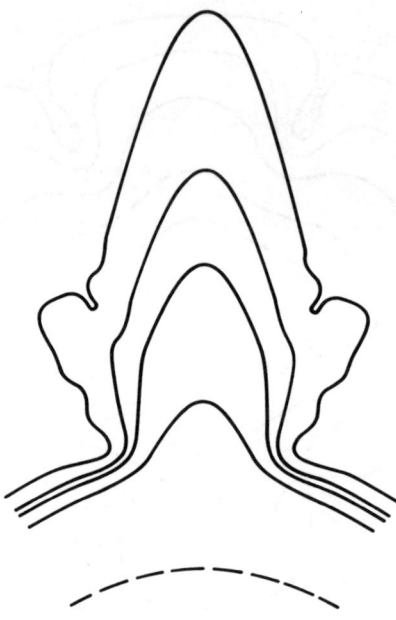

Fig. 3 Sequence of solidification fronts (unequal time intervals) generated by the boundary-layer model with a six-fold anisotropy in the growth kinetics. The initial state (dashed curve) is an unperturbed circle.

initial configurations. The pair of large side branches visible in the figure is reminiscent of structures seen in real snowflakes (Nakaya, 1954) and is apparently a transient phenomenon generated by the initial instability at the hexagonal corners. The parabolic tips do seem to be emitting weak side-branching oscillations which are best seen in graphs of curvature *versus* arc length; but we have not been able to continue the computation long enough to discover whether or not these are also transients as in the geometrical model.

By far the most stunning surprise to emerge so far from work on the boundary-layer model and the geometrical model as well has been the discovery that the velocities and tip radii of dendrites selected by dynamical simulations like those just described can be predicted by examining the solvability of the steady-state equations (Ben-Jacob *et al.*, 1984a). In our first paper on the boundary-layer model (Ben-Jacob *et al.*, 1983; 1984b), we pointed out that surface tension is a singular perturbation, and that the continuous family of Ivantsov needle crystals must break down when d_0 is nonzero. The trouble is that the shape of the tip, which is no longer parabolic if there are capillary corrections, cannot necessarily be matched smoothly onto a parabola behind it except possibly for special values of the velocity. My initial interpretation of this situation was that dendritic velocities had to be selected by dynamical properties of the tip, and that consistency with steady-state parabolic solutions far behind the tip had to be irrelevant. Real dendrites, with side branches, simply do not look like that. It turns out, however, that whenever we have found a propagating tip in a dynamical simulation, the velocity and radius of curvature of that tip have had precisely those special values for which there exist needle-crystal solutions of the steady-state equations. Side branching dendrites without an associated needle crystal might occur in this model; but so far we have not seen such behavior.

The picture that now seems to us to be emerging from ongoing analytic and numerical work is the following. Think of the equations of motion for the boundary-layer model as a dynamical system which describes trajectories in a space of functions $[\kappa(s, t), h(s, t)]$. Depending on the values of the system parameters Δ, $d_0(\theta)$, $\beta_0(\theta)$, this space may (or may not) contain stationary states, that is, dynamical fixed points $[\kappa^*(s), h^*(s)]$ which describe needle crystals and which satisfy

$$\left(\frac{\partial \kappa^*}{\partial t}\right)_s = \left(\frac{\partial h^*}{\partial t}\right)_s = 0 \ . \tag{4.9}$$

A special property of the boundary-layer model is that these fixed points, if they exist at all, can be no more stable than marginally stable. One can show analytically that there always exist modes of deformation of a needle crystal which relax infinitely slowly. Physically, these modes are a result of the slow dynamics of the boundary layer which becomes indefinitely thick as one moves away from the tip. (No such dynamical feature occurs in the geometrical model.) If more than one fixed point exists, the most nearly stable one will be that with the largest velocity and smallest tip radius; slower, flatter dendritic tips are more likely to be unstable against tip splitting deformations. It can happen that even the fixed point with largest velocity is strongly unstable, for example, if the crystalline anisotropy is too weak. In that case, no dendritic pattern formation occurs, and the morphology probably becomes increasingly irregular as the system evolves. It might also happen that the leading fixed point is stable, that even the marginal trajectories in $[\kappa, h]$ space are attracted to it but approach algebraically rather than exponentially. Then the boundary-layer model would behave like the geometrical model (Kessler *et al.*, 1984) and produce propagating needles with only transient or fluctuation-induced side branching activity. It is possible that this is a correct physical picture, that side branching occurs because dendritic tips have just enough instability to amplify special fluctuations in their environments.

The most interesting possibility is that the leading fixed point is weakly, or perhaps just marginally, unstable. I think that this also could correspond to physical reality. Real dendrites are not needle-crystal fixed points but, rather, might be visualized as limit cycles in $[\kappa, h]$ space. Near the tip, the dendritic shape may repeat itself with a period equal to the interval between emission of side branches. [By using the κ representation, we consider only the shape of the interface and avoid problems associated with translational motion in real space. Note the time derivatives at fixed s in (4.9).] Irregular coarsening of side branches behind the tip means that this limit cycle would have to be weakly unstable; but that instability could have little to do with the velocity-selection mechanism. The apparent role of the fixed point would be to serve as an almost stable center for the limit cycle. All small perturbations of the needle crystal, that is, all dynamical trajectories starting near the fixed point, would be attracted onto the limit cycle. For this to happen, the fixed point must be almost but not quite stable, a situation which looks something like marginal instability, but which I do not know how to define precisely.

In each of these dynamical pictures of dendrites, the solvability condition serves as a selection principle because the fixed point possesses some average properties of the nearby dynamical trajectory. In order to use this scheme to make practical calculations, one need not solve the full dynamical problem. One simply solves the steady-state equations, which can now be thought of as a nonlinear eigenvalue problem for the velocity, and checks to see whether that solution is sufficiently (marginally?) stable. There remains some vestige of the earlier marginal-stability hypothesis (Langer, 1980; Langer and Müller-Krumbhaar, 1978) here, along with the hint of a connection between marginal stability and the sensitivity of pattern forming systems to small changes in growth conditions. But the picture is clearly very different from anything that has been proposed previously.

I remarked at the beginning of this section that the main purpose of the boundary-layer model — and the geometrical model as well — was to provide new insight into mathematical mechanisms of pattern selection. The investigations described above certainly have produced new ideas; but it remains to be seen whether these ideas are relevant to reality as defined by the basic model in Sec. 2. A crucial question is whether the breakdown of the Ivantsov solutions, and therefore the existence of a solvability condition, might be a special feature of the local contour-dynamical models. It is conceivable that the fully nonlocal basic model behaves quite differently, that the continuous family of Ivantsov solutions survives the effects of capillarity, and that some more complex dynamical mechanism is responsible for dendritic pattern selection. This question is being studied actively by a number of people (including this author) as this essay is being written. No results are certain yet, but there are strong indications in favor of a solvability condition for the fully nonlocal model. If this result is confirmed, then we shall be on our way toward fitting one small piece into a very large puzzle.

Acknowledgments

This research was supported in part by the U.S. Department of Energy Grant No. DE-FG03-84ER45108 and by the National Science Foundation under Grant No. PHY 82-17853, supplemented by funds from the National Aeronautics and Space Administration, at the University of California at Santa Barbara.

References

Ahlers, G., Cannell, D. S. and Steinberg, V. (1985) *Phys. Rev. Lett.* **54**, 1373.

Belintsev, B. N. (1983a) *Usp. Fiz. Nauk* **141**, 55.

Belintsev, B. N. (1983b) *Sov. Phys. Usp.* **26** (9), 775.

Ben-Jacob, E., Goldenfeld, N. D., Langer, J. S. and Schön, G. (1983) *Phys. Rev. Lett.* **51**, 1930.

Ben-Jacob, E., Goldenfeld, N. D., Kotliar, B. G. and Langer, J. S. (1984a) *Phys. Rev. Lett.* **53**, 2110.

Ben-Jacob, E., Goldenfeld, N. D., Langer, J. S. and Schön, G. (1984b) *Phys. Rev.* **A29**, 330.

Ben-Jacob, E., Godbey, R., Goldenfeld, N. D., Koplik, J., Levine, H., Mueller, T. and Sander, L. M. (1986) *Phys. Rev. Lett.* (to be published).

Bialek, W. S. (1983) "Quantum Effects in the Dynamics of Biological Systems," Ph.D. thesis, University of California, Berkeley (unpublished).

Brower, R., Kessler, D., Koplik, J. and Levine, H. (1983) *Phys. Rev. Lett.* **51**, 1111.

Brower, R., Kessler, D., Koplik, J. and Levine, H. (1984) *Phys. Rev.* **A29**, 1335.

Cahn, J. W. and Hilliard, J. E. (1958) *J. Chem. Phys.* **28**, 258.

Chalmers, B. (1964) *Principles of Solidification* (Wiley, New York).

Collins, J. B. and Levine, H. (1985) *Phys. Rev.* **B31**, 6119.

Deutsch, J. M. (1985) (unpublished).

Fix, G. (1983) "Free Boundary Problems, Theory and Applications," *Research Notes in Mathematics*, Vol. 2, Fasans, A. and Primicero, M., eds. (Pitman, New York).

Glicksman, M. E., Shaefer, R. J. and Ayers, J. D. (1976) *Metall. Trans.* **A7**, 1747.

Halperin, B. I., Hohenberg, P. C. and Ma, S.-k. (1974) *Phys. Rev.* **B10**, 139.

Hele-Shaw, H. S. S. (1898) *Nature* **58**, 34.

Heutmaker, M. S., Fraenkel, P. N. and Gollub, J. P. (1985) *Phys. Rev. Lett.* **54**, 1369.

Hobbs, P. V. (1974) *Ice Physics* (Clarendon, Oxford).

Hohenberg, P. C. and Halperin, B. I. (1977) *Rev. Mod. Phys.* **49**, 435.

Horvay, G. and Cahn, J. W. (1961) *Acta Metall.* **9**, 695.

Huang, S. C. and Glicksman, M. E. (1981) *Acta Metall.* **29**, 701, 717.

Ivantsov, G. P. (1947) *Dokl. Akad. Nauk* SSSR **58**, 567.

Kessler, D., Koplik, J. and Levine, H. (1984) *Phys. Rev.* **A30**, 3161.

Langer, J. S. (1980) *Rev. Mod. Phys.* **52**, 1.

Langer, J. S. and Müller-Krumbhaar, H. (1978) *Acta Metall.* **26**, 1681, 1689, 1697.

Nakaya, V. (1954) *Snow Crystals* (Harvard University Press, Cambridge, Massachusetts).

Oldfield, W. (1973) *Mater. Sci. Engr.* **11**, 211.

Saffman, P. G. and Taylor, G. I. (1958) *Proc. Roy. Soc.* **A245**, 312.

Smith, J. B. (1981) *J. Comp. Phys.* **39**, 112.

Thompson, D. W. (1944) *On Growth and Form* (MacMillan, New York).

Tryggvason, G. and Aref. H. (1983) *J. Fluid. Mech.* **136**, 1.

Whitehead, J. A., Jr. (1975) "A Survey of Hydrodynamic Instabilities" in *Fluctuations, Instabilities and Phase Transitions,* Riste, T., ed. (Plenum, New York).

Witten, T. A. and Sander, L. M. (1983) *Phys. Rev.* **B27**, 5686.

Woodruff, D. P. (1973) *The Liquid-Solid Interface* (Cambridge University Press, Cambridge, England).

QUANTUM MECHANICS AT THE MACROSCOPIC LEVEL

Anthony J. Leggett

Department of Physics
University of Illinois
1110, W. Green Street
Urbana, IL 61801
USA

1. Introduction

We are all familiar with the idea that while the microscopic description of the physical world requires quantum mechanics, at the macroscopic level a classical description suffices. What, exactly, does this mean? It certainly does *not* mean that quantum mechanics plays no role in the physics of macroscopic bodies; quite the contrary, all our detailed understanding of the behavior of matter in bulk is firmly based on quantum-mechanical principles. Nor does it mean that characteristically quantum-mechanical effects cannot be observed over macroscopic length (and time) scales, or when macroscopic numbers of particles are involved; a spectacular counter-example is the beautiful experiment (Jaklevic *et al.*, 1965) which demonstrates the interference properties of the superconducting wave function in a superconducting ring interrupted by two Josephson junctions.

What is meant is rather the following. In our ordinary, everyday language we talk of things happening or not happening: chairs are in this room or the next, counters click or do not click, and so on. (This is so obvious and universal a feature of our way of describing the world that it would not normally occur to us to comment on it.) Classical physics, for all its departures from "common-sense" ideas in other directions, preserves this feature: whether the variable we are talking about is the position of a particle, (at a given time), the value of a field (at a given space-time point) or something else, it is at least consistent to suppose that this variable does "actually have" a given value, whether or not we are in a position to know it. In other words, classical physics is a "realistic" theory in the philosophical sense. By contrast, quantum mechanics in its standard textbook formulation talks in terms of probability amplitudes, and denies that a microscopic entity such as an electron always possesses definite properties. For example, if a particular atom is drawn from an ensemble described, as regards its spin behavior, by the wave function

$$\psi = a\psi_+ + b\psi_- \tag{1.1}$$

(where ψ_+, ψ_- denote respectively the states of spin $\pm \frac{1}{2}\hbar$ along some chosen axis) then, even though measurement of the spin will reveal one of the values $+\frac{1}{2}\hbar$, $-\frac{1}{2}\hbar$, we are not supposed to infer that the spin of the atom in question "actually had" one of these values (or indeed any value at all) before the measurement was made. In fact, the standard interpretation of quantum mechanics tells us very firmly that the concept of a microscopic entity such as an atom "having" definite properties in the absence of a specific experimental setup designed to measure these properties is in general quite meaningless (Bohr, 1963). At the microscopic level, quantum mechanics is *not* a "realistic" theory.

What is meant by the above claim, then, is that when one is interested in describing the world at the macroscopic level (the level of chairs and counters, let us say) and provided one is interested only in macroscopic properties, then the quantum mechanical principle of superposition is irrelevant: to all intents and purposes we can say that macroscopic properties do take definite values, subject perhaps to some uncertainty or indeterminacy which is microscopic and hence negligibly small compared to the quantities themselves. This claim is, of course, in no way in conflict with the realization that the actual values of the macroscopic properties may well be determined by effects at the microscopic level which are intrinsically quantum in nature; for example, the fact that the mean thermal energy of a metal at low temperatures is proportional to T^2 is a consequence of the quantum mechanical behavior of the electrons, but this does not mean that we need to consider superpositions of states of the metal with widely different energies. Similar remarks apply to the superconducting interference experiment mentioned above: although the superconducting wave function (wave function of the center of mass of the Cooper pairs) is a quantity whose physical meaning can only be explained in quantum-mechanical terms, in interpreting the experimental results it is perfectly adequate to regard it as a classical field whose value is well-defined at each space-time point (more on this below). In sum, it is claimed that at the macroscopic level, whatever the quantum origins of the actual values of macroscopic properties, or even the laws which they obey, in describing these properties the *language* of classical physics (i.e. of realism) is adequate; in particular, we never have to deal with states in which a macroscopic variable is not even approximately defined, i.e. a quantum superposition of macroscopically different states.

It is quite impossible to overestimate the significance of this claim for the interpretation of our modern quantum-mechanics-based scheme of the world. Indeed, it is the very cornerstone of the Copenhagen interpretation, at least, in the version espoused by Niels Bohr (Bohr, 1963). In this version it is said that a microscopic entity, such as an atom, is simply not the *kind of thing* to which it makes sense to ascribe properties in the absence of a specification of the macroscopic instruments with which it can interact, — it is, as it were, only a link between the macroscopic preparation device and the macroscopic measuring instrument — and Bohr recognized very clearly that this interpretation is only tenable if the metaphysical status of the macroscopic instruments themselves, and their properties, is not in doubt. In fact, he repeatedly emphasized that the preparation and measuring devices themselves *must* be described in the language of classical physics (meaning the language of realism); if this were not done, he pointed out, we should have no way of communicating our experimental procedure and our results unambiguously to one another. Thus, the qualitative distinction between the microscopic level (where a realistic interpretation is forbidden) and the macroscopic one (where it is essential) is absolutely central to all of Bohr's thinking,[1] and to much of the vast literature on the interpretation of quantum mechanics over the last sixty years. Such a distinction is only tenable, of course, if the claim described above is correct.

Now, there are at least two alternative reasons why the claim could be correct. The first possibility is that quantum mechanics is simply not the right theory to describe the world at the macroscopic level — that qualitative changes in the basic laws of physics occur when matter becomes sufficiently complex. Any such proposal would of course have to take careful account of the fact that many, indeed most, of the "gross" macroscopic properties of matter are well accounted for by microscopic quantum-mechanical considerations; nevertheless, a scheme in which this good agreement is preserved, but macroscopic realism is automatically achieved, is by no means excluded by any existing experiment. I have explored this possibility at some length elsewhere (Leggett, 1980, 1986) and will not discuss it further here; at present, probably only a small minority of physicists take it seriously.

The alternative possibility is that quantum mechanics is indeed in principle a complete and universal theory of the world, including the macroscopic properties of macroscropic bodies, but that consistent application of the formalism leads to the result indicated, i.e., that the macroscopic properties may be adequately described in a classical, realistic language, and in particular that quantum superpositions of macroscopically different states never occur in real life. Much of the literature (de Witt and Graham, 1971) on the quantum theory of measurement has in fact been devoted precisely to an attempt to establish this conclusion; for an extended recent discussion, see e.g. Joos and Zeh (1985). Crudely speaking, one can identify two main themes in these attempts. The first is that, simply because the macroscopic properties of a macroscopic body (such as the position or velocity of the center of mass) are sums or averages of very many ($\sim N$) microscopic quantities, the usual statistical considerations ensure that the *relative* uncertainty of these quantities will be very small ($\sim N^{-1/2}$, usually). For example, consider a set of N noninteracting particles each of mass m placed in a potential $\frac{1}{2} m \omega_0^2 (x - x_0)^2$, where x_0 is some fixed distance. In the quantum-mechanical ground state of this system, the root-mean-square displacement of each particle from its mean position x_0 is given by $\sqrt{\hbar / 2 m \omega_0}$, a quantity which we may suppose for the sake of the argument to be comparable to x_0. However, in the case of the center-of-mass

---

[1] Though cf. Jammer (1974) and Leggett (1985).

coordinate $X \equiv N^{-1} \sum_i x_i$ we have $\langle X \rangle = x_0$, but

$$\langle X^2 \rangle = N^{-2} \left\langle \left(\sum_i x_i \right)^2 \right\rangle = N^{-2} \sum_{ij} \langle x_i x_j \rangle \quad . \tag{1.2}$$

Because the different particles are uncorrelated, we have $\langle x_i x_j \rangle = \langle x_i \rangle \langle x_j \rangle + \delta_{ij} (\Delta x_i)^2 = x_0^2 + \delta_{ij} (\sqrt{\hbar/2m\omega_0})^2$, and hence

$$\langle X^2 \rangle = x_0^2 + N^{-1} (\hbar/2m\omega_0) \quad ,$$

$$\Delta X \equiv \sqrt{\langle X^2 \rangle - \langle X \rangle^2} = \sqrt{\langle X^2 \rangle - x_0^2} = N^{-1/2} \sqrt{\hbar/2m\omega_0} \tag{1.3}$$

which for large enough N can be made arbitrarily small compared to the average position x_0. It is amusing to note that (1.3) is identical to the result which would follow if the potential acted on the center of mass alone and was of the form $\frac{1}{2} M \omega_0^2 (X - x_0)^2$; in this case the result (1.3) would be quite independent of whether or not there were interactions between the constituent particles.

Arguments similar to the above one (which can be found in many elementary textbooks on quantum mechanics) do indeed make it plausible that in most real-life situations the quantum-mechanical uncertainty in the value of a macroscopic variable is negligibly small compared to its typical mean values (though cf. below, Sec. 2), and hence that a classical description is applicable. However, there is one type of situation, which we may call the "Schrödinger's Cat" (Schrödinger, 1935) type, which they spectacularly fail to cover. Such a situation occurs when the value of a microscopic variable (such as the spin of an atom, or the polarization of a photon) is used to trigger a macroscopic event whose nature depends on its value, and where the microscopic system in question starts in a linear superposition of two or more states which correspond to different values of the "triggering" variables. To illustrate this, let us consider the following, admittedly rather artificial, thought-experiment (see Fig. 1). A macroscopic billiard-ball is placed in a shallow hollow close to the edge of much deeper potential well, in such a way that if struck by a single sufficiently energetic photon its recoil momentum on absorbing the photon will be sufficient to take it over the low barrier and set it rolling into the deep well.[2] A light source, and a set of polarizers and mirrors, is then set up in such a way that light emitted by the source will eventually strike the billiard ball, but that its

---

[2] A proper quantum-mechanical calculation would have to take account of the fact that the laws of conservation of momentum and energy (*inter alia*) in general require not only recoil of the billiard ball's center of mass but also some excitation of its internal degrees of freedom. Since this example is only provided to motivate the main work of this paper, I shall not discuss this complication here: it does not affect the essence of the argument.

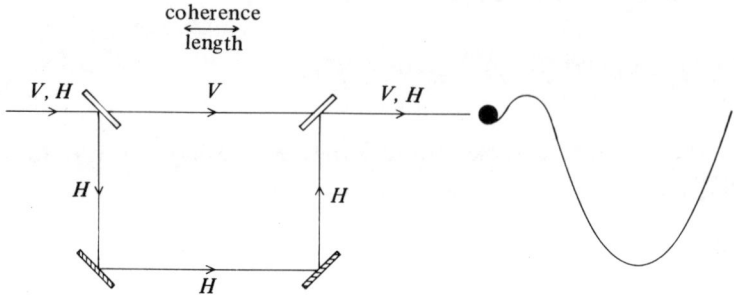

Fig. 1. Schematic device for the amplification of microscopic effects. V and H indicate the paths of vertically and horizontally polarized photons respectively.

time of transit through the apparatus will depend on its polarization. Let the time of transit for light which was originally polarized in a vertical plane be t_0, that of horizontally polarized light $t_0 + \Delta t$, and assume that both t_0 and Δt are large compared to the typical coherence time of the source. Imagine now that the source emits a single photon at time $-t_0$. If the photon is vertically polarized (corresponding to wave function ϕ_1) it will strike the billiard ball at time 0, and the latter will absorb it, mount the barrier and start rolling down into the deep well at this time. Let the (time-dependent) wave function which describes the motion of its center of mass be $\Psi_1(t)$. If on the other hand the photon is horizontally polarized (wave function ϕ_2), it will not strike the billiard ball until time Δt, and as the latter rolls down into the deep well its wave function will be $\Psi_1(t - \Delta t) \equiv \Psi_2(t)$. With suitable adjustments to the parameters we can ensure that, while in either of the states Ψ_1, Ψ_2 separately the quantum-mechanical uncertainty in the value of the center-of-mass co-ordinate X at a given time t_0 is small compared to its typical mean value in that state,[3] the *difference* between the mean value of $X(t)$ in state Ψ_1 and Ψ_2 is as large as we please; thus, $\Psi_1(t)$ and $\Psi_2(t)$ clearly correspond to macro-scopically different states.

Now suppose that the photon emitted at time $-t_0$ was neither horizontally nor vertically polarized but rather was in a linear superposition of the two states (corresponding classically to light polarized at some angle other than 0 or $\pi/2$ to the vertical): that is, its wave function was

$$\phi(t) = \alpha\phi_1(t) + \beta\phi_2(t) , \qquad |\alpha|^2 + |\beta|^2 = 1 . \tag{1.4}$$

---

[3]This value corresponds, roughly, to the position at time t of a *classical* billiard-ball which was struck in the way indicated.

Because of the strict linearity of the laws of quantum mechanics, we can say without detailed argument that in this case the final state of billiard-ball will be represented by the corresponding linear superposition of Ψ_1 and Ψ_2 :

$$\Psi(t) = \alpha\Psi_1(t) + \beta\Psi_2(t) \quad . \tag{1.5}$$

Thus, *prima facie*, we have induced in the billiard ball, a macroscopic object, a quantum-mechanical superposition of states corresponding to macroscopically different properties, in contradiction to the claim stated above.

On the whole, discussion of states of the form (1.5) in the literature has tended not to focus on the technical difficulty of producing them (which is in fact considerable). Rather, it has tended to emphasize the point that even if such states can be produced, it will (allegedly) be impossible to distinguish them from a classical mixture of the states Ψ_1 and Ψ_2, i.e. a description in which the macroscopic system is known to be either in state 1 or in state 2, and these two possibilities are assigned a probability of $|\alpha|^2$ and $|\beta|^2$ respectively. How, in principle, could we distinguish the two descriptions experimentally? Consider the measurement of the expectation value of some operator $\hat{\Omega}$ at time t. Evidently this will be given, for the mixture, by the expression

$$\langle\Omega\rangle_{\text{mixt}} = |\alpha|^2 \langle\Psi_1(t)|\hat{\Omega}|\Psi_1(t)\rangle + |\beta|^2 \langle\Psi_2(t)|\hat{\Omega}|\Psi_2(t)\rangle \quad . \tag{1.6}$$

For the quantum superposition (pure state) (1.5), on the other hand, we have

$$\langle\Omega\rangle_{\text{pure state}} = \langle\Psi(t)|\hat{\Omega}|\Psi(t)\rangle = |\alpha|^2 \langle\Psi_1(t)|\hat{\Omega}|\Psi_1(t)\rangle$$
$$+ |\beta|^2 \langle\Psi_2(t)|\hat{\Omega}|\Psi_2(t)\rangle + \text{Re}\{\alpha^*\beta \langle\Psi_1(t)|\hat{\Omega}|\Psi_2(t)\rangle\}. \tag{1.7}$$

Thus it is the last term in (1.7), the so-called *interference-term*, which distinguishes the two descriptions. The alleged impossibility of making the distinction is therefore often expressed by saying that "it is impossible to see quantum interference between macroscopically distinct states."

Once this claim is granted, the argument then usually proceeds by claiming that since there is no measurement which could possibly distinguish the pure state (1.5) from the corresponding mixture, the macroscopic system "in effect" *is* in a mixture; the "in effect" is then slyly dropped, and one ends up with the claim that macroscopic objects are always in definite macroscopic states just as "common sense" would have us assume. This second step seems to the present author, and some others, (see e.g., d'Espagnat (1976), Bub (1968)) to embody a severe logical fallacy, and indeed most of the more careful writers on the quantum theory of measurement have recognized it as, at least, problematical; in any case, the issues involved at this stage are not of the kind which could be resolved by a more careful scrutiny of the application of the quantum formalism itself. For the purposes of the present paper, therefore, I will not discuss it further but will concentrate on state I of the argument, namely the claim that interference effects always vanish at the macroscopic level.

The arguments which are advanced in the literature for this conclusion are formally many and varied, but one can probably isolate two main themes, or rather perhaps two variants of a single theme. In the first place it is claimed that even in the case where the relevant wave functions Ψ_1 and Ψ_2 can be written as functions only of a single macroscopic variable (e.g., in the above example, the center-of-mass coordinate of the billiard ball) it may be impossible in practice to construct a measurable operator $\hat{\Omega}$ which has nonzero matrix elements between macroscopically different states. Secondly — and more importantly for the purposes of this paper — it is argued that a macroscopic object is inevitably coupled so strongly to its environment that, even if we could initially prepare a state of type in Eq. (1.5), with the "environment" in some definite state, say $\chi(t)$, then the time evolution of the coupled system and environment would very rapidly convert the wave function of the whole into the form

$$\Psi(t) = \alpha\chi_1(t)\Psi_1(t) + \beta\chi_2(t)\Psi_2(t) \tag{1.8}$$

where χ_1 and χ_2 are mutually orthogonal states of the environment. Once the properties of the system and its environment have become entangled in this way, it follows from a trivially demonstrable theorem of quantum measurement theory (see e.g. Leggett (1980)) that no measurement on the system alone can distinguish the state (1.8) from a classical mixture of the states $\chi_1\Psi_1$ and $\chi_2\Psi_2$. Any attempt to distinguish the two descriptions would have to involve, in effect, a complete measurement of the properties of the environment — and since the latter can include, for example, photons radiated away to infinity, any such measurement would be totally out of the question.

The above crude summary cannot of course do justice to the detailed and subtle arguments which have been developed over the last sixty years by the practitioners of the quantum theory of measurement. As a result of these arguments, it is probably fair to say that there is a high degree of consensus, among those physicists who have thought seriously about the question, that it will always be impossible to see quantum interference between macroscopically different states. In the rest of this paper I shall describe some developments of the last five years or so in solid-state physics which imply that *this conclusion is very probably incorrect.*

Two remarks need to be made at once. First, there is absolutely nothing wrong in principle with the general arguments which have been developed in quantum measurement theory. In particular, it is trivially possible to make the claim of vanishing interference at the macroscopic level correct by defining "macroscopic" to mean "in the limit $N \to \infty$" where N is the number of particles involved (cf., e.g., Hepp (1972) and Anderson (1984)). But taking "the limit $N \to \infty$", is merely a mathematical device which happens to be useful in certain technical contexts; in real life, objects which we would legitimately regard as macroscopic, in particular ourselves and our measuring instruments, correspond to large but finite values of N, and it is clear that if it is to be used in the context, for example, of the Copenhagen interpretation it is at this level that the claim must hold. We shall, in fact, have to define "macroscopic" in a rather looser way (though one which is, I believe, still consistent with the everyday use of the term) in order to provide a plausible counterexample: our prime candidate system will in fact be a superconducting device, containing of order 10^{23} electrons, in which the "macroscopically different" states whose quantum interference we seek correspond to states in which a current of the order of 1 microamp is flowing clockwise and counterclockwise respectively. While this level is no doubt rather below that of "everyday life", it is a very great way above that of electrons, atoms and molecules — the "quantum microworld" of Bohr — and it is in fact extremely remarkable that there is even a reasonable chance of seeing quantum interference in such systems. In fact, the only reason why the usual considerations of the quantum theory of measurement do not totally knock out the possibility is a remarkable concatenation of favorable factors, associated with recent advances in microfabrication, cryogenic and noise control technologies. *A priori*, it would have been highly optimistic to suppose that these factors would come together in any system which could reasonably be called macroscopic; but in this instance Nature seems to have been kind to us.

The second remark which is necessary is that when one says that there is a reasonable hope of seeing quantum interference at the macroscopic level, one is of course making the implicit assumption that quantum mechanics is in principle a complete and universal theory of the physical world. As pointed out above, this is not the only possible assumption. Indeed, if we can reach the parameter regime where the theory to be described below (which is, of course, entirely quantum-mechanical) predicts that interference effects should be seen, and if the experiments then fail to show such interference, then this would provide serious evidence against the universality of the quantum description.

The rest of this paper is devoted to a review of our current understanding of the quantum mechanics of a macroscopic variable, bearing in mind the ultimate goal of achieving a situation where not only is that variable described by a quantum-mechanical wave function of the general type (1.5) but where we can actually verify (or not!) the existence of the interference terms. To explain the plan of the paper, it may be easiest to start at the end. It turns out that the most promising situation in which to look for incontrovertible evidence of the existence of the quantum interference terms is one where the motion of the macroscopic variable in question is governed by a potential energy with two degenerate wells (see Fig. 2b). This situation is closely analogous to that of the nitrogen atom in the NH_3 (ammonia) molecule, and just as in that case we would *prima facie* expect the system to tunnel between the two wells in an oscillatory fashion. As we shall demonstrate below, observation of this phenomena would provide striking confirmation of the hypothesis that even at the macroscopic level Nature does not believe in realism. However, in order to be sure of this prediction we should first confirm that we expect the macroscopic variable in question to be able to tunnel through potential barriers at all, and it turns out that it is much easier to check this assumption experimentally in a "decay" situation (Fig. 2a), that is where the system can escape from a single metastable well into infinite space (as in the decay of a nucleus by alpha-particle emission). Moreover, it is crucial in the discussion of both phenomena to take into account all the considerations developed in the quantum theory of measurement, and in particular to make sure that our model adequately recognizes the strong coupling to its environment which is characteristic of a macroscopic

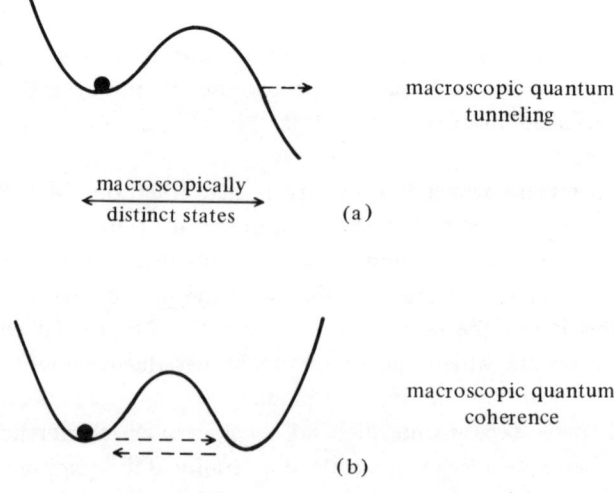

Fig. 2. Macroscopic quantum tunneling and coherence.

system. Finally, we need to select one or more specific physical systems as particularly appropriate to our purpose. It turns out that a particularly appropriate class of solid-state systems is that of superconducting devices based on the Josephson effect, and much of the relevant work has been done specifically in the context of such devices (though cf. also e.g., Bardeen (1979)); thus, in the present paper I will focus on these systems whenever I need to illustrate the considerations by a specific model. Most of the general conclusions are in no way restricted to these systems.

In the following sections I take the above points in the reverse order. First, I give a brief discussion of the superconducting devices which are the prime candidate for our ultimate goal (Sec. 2). Next, I consider the general question of the quantum mechanics of a macroscopic variable, taking into account its interaction with its environment (Sec. 3). In Sec. 4 I examine the question of the "decay" of a state in which the macroscopic variable is trapped in a meta-stable potential well (a phenomenon known in the literature as macroscopic quantum tunneling or MQT, Fig. 2a), and compare the theoretical predictions for this phenomenon with the experiments which have so far been carried out on superconducting devices. Finally, in Sec. 5 I come to the experiment on "macroscopic quantum coherence" (MQC, Fig. 2b) (the analog of the NH_3 experiment) which would, if carried out successfully, be the most dramatic and direct evidence for quantum interference at the macroscopic level.

My main aim in this paper is to give the reader a feeling for what we currently think we do and do not understand in this field. I have therefore concentrated on the conceptual issues, and have usually referred to the rapidly growing literature for the technical details of the calculations whose results I quote. In particular, most of the calculations referred to in Sec. 2, in Secs. 3 and 4 (except for the finite-temperature results) and in Sec. 5 may be found in Leggett (1984b), Caldeira and Leggett (1983) and Leggett *et al.* (1985) respectively.

2. Superconducting Devices

In planning tests of the quantum-mechanical superposition principle at the macroscopic level, a number of desiderata for the physical system which we will use must be borne in mind. First, is should possess two or more discrete states which by a reasonable "common-sense" criterion can be called macroscopically distinct. Secondly, the transition probability between these states should nevertheless not be negligibly small; this is possible if the energy barrrier which separates them is itself of a microscopic rather than a macroscopic order of magnitude. Thirdly, the irreversible coupling between the variable describing the macroscopic properties and the myriad other degrees of freedom of the system should be sufficiently weak that it does not necessarily wash out the quantum coherence properties. Fourthly, we should have either a microscopic model of the system in which we can put complete confidence, or failing that at least a comprehensive and generally agreed description of its *classical* behavior. Finally, we need to be able to vary the control parameters sufficiently both to check our microscopic model and/or our classical description, and to make a range of predictions about the macroscopic quantum behavior.

While there exist a number of different systems which satisfy some of the above criteria, e.g. charge density waves in quasi-one-dimensional systems (Bardeen, 1979) and the photon field in a bistable ring laser (Lett *et al.*, 1981), the only class of systems which at present comes close to satisfying all of them seems to be superconducting devices based on the Josephson effect, and I shall therefore particularize the discussion to these systems. From a conceptual standpoint the simplest such system is a thick (cf. below) superconducting ring interrupted by a Josephson junction; while such junctions come in many different varieties, for our purposes they may be thought of provisionally as constrictions or barriers through which the electrons of the bulk metal can pass, but with some difficulty. The system just described is frequently referred to for brevity as an "rf SQUID" (radio frequency superconducting quantum interference device), since it is the active element in the latter device. For a description of the principal features of such devices and their use as magneto-meters, see Lounasmaa (1974). An essential feature of this system is that it is possible to apply a variable *external* magnetic flux through the ring; the system will then in general respond by exciting dc circulating currents which will generate additional magnetic flux. We will assume that the externally applied flux can be treated as a classical control parameter; the internally generated flux, and hence the total flux trapped in the ring, will however be treated as a quantum-mechanical variable. A system closely related to the rf SQUID is obtained by breaking the ring open and driving through it (i.e. through the Josephson junction) a fixed external current; this system is usually called a "current-biased junction". Because this system, in distinction to the rf SQUID, involves the exchange of energy with an external source (the battery driving the current), the analysis of it involves some conceptual problems which are absent in the former case. Hence I will mainly discuss the rf SQUID, and comment where necessary on the differences which arise in the analysis for a current-biased junction.

Our modern understanding of the phenomenon of superconductivity, both in the bulk and in Josephson devices, is based on the idea (Bardeen *et al.* (1957)) that electrons near the Fermi surface form Cooper pairs, which automatically then undergo a sort of Bose condensation. The pairs in effect have a mutual interaction energy, and moreover, because of their charge, interact with both externally and internally generated electric and magnetic fields. For the sake of clarity of exposition, let us begin with a very simple system — uncharged noninteracting bosons — and add the complications one by one.

Suppose, first, that our N bosons were confined to a thick uninterrupted ring with uniform cross-section. The energy eigenstates of a single particle in such a ring are of the general form

$$\phi(\mathbf{r}) = \chi(\mathbf{r}_\perp) \exp(im\theta) \quad , \tag{2.1}$$

where $\mathbf{r}_\perp$ schematically denotes the coordinate in the cross-sectional plane and θ is the angle around the ring (cf. Fig. 3). To ensure single-valuedness of the wave function we must require m to be an integer (including zero). Because the Hamiltonian at this stage is invariant under time reversal, the pair of states corresponding to m and $-m$ are degenerate, so a more general set of energy eigenfunctions would be the functions of the form

$$\phi(\mathbf{r}) = \chi(\mathbf{r}_\perp)(ae^{im\theta} + be^{-im\theta}); \qquad |a|^2 + |b|^2 = 1 \quad . \tag{2.2}$$

Note that while for a single particle described by the wave function (2.1) the probability density is uniform around the ring, for one described by (2.2) it is in general strongly inhomogeneous. Since the N bosons are at this stage non-interacting, an energy eigenfunction of the total system can be built up by putting each particle in a state of the form (2.2), with a particular value of

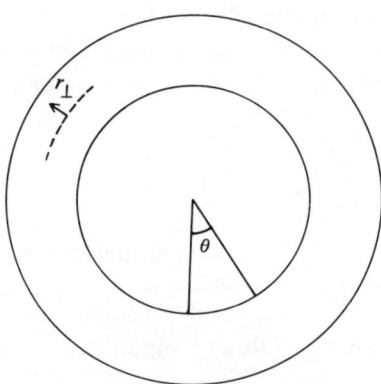

Fig. 3. A bulk superconducting ring. The ring is doughnut-shaped and $\mathbf{r}_\perp$ measures the position of a given point in the cross-sectional plane, as indicated.

a and b, and symmetrizing the whole, that is,[4]

$$\Psi(\mathbf{r}_1\,\mathbf{r}_2\ldots\mathbf{r}_N) = S \prod_{i=1}^{N} \chi(\mathbf{r}_{\perp i})\,\{a_i\exp(im_i\theta_i) + b_i\exp(-im_i\theta_i)\}\ .$$

(2.3)

The most general energy eigenfunction is a linear superposition of all states of the form (2.3) which corresponds to the same total energy. Two special (extreme) cases are of particular interest.

(A) $\Psi(\mathbf{r}_1\,\mathbf{r}_2\ldots\mathbf{r}_N) = \prod_{i=1}^{N} \chi(\mathbf{r}_{i\perp})\,\{a\exp(im\theta_i) + b\exp(-im\theta_i)\}\ .$ (2.4)

This is a special case of (2.3) and indicates a state in which all the N bosons are condensed into a single one-particle state. It does *not* correspond to a super-position of macroscopically distinguishable states.

(B) $\Psi(\mathbf{r}_1\,\mathbf{r}_2\ldots\mathbf{r}_N) = \prod_{i=1}^{N} \chi(\mathbf{r}_{i\perp})\left\{A \prod_{i=1}^{N} \exp(im\theta_i) + B \prod_{i=1}^{N} \exp(-im\theta_i)\right\}\ .$

(2.5)

This can be written as a linear superposition of states of the form (2.3), and corresponds to a superposition of macroscopically distinguishable states with angular momentum Nm and $-Nm$ respectively. For the noninteracting gas of boson (A) and (B) are both energy eigenstates, with the same energy.

Next, we switch on some short-range interactions between the constituent bosons. To ensure stability we make this interaction repulsive. (The resultant model is in fact a crude model for real liquid ^{4}He.) Now, in the crudest approxi-mation the main effect of the repulsion will be to increase the energy of states which have an inhomogeneous particle density over those where the density is homogeneous. It is easily verified that the expectation value of the total particle density

$$\rho(\mathbf{r}) \equiv \sum_i \delta(\mathbf{r}-\mathbf{r}_i)$$

(2.6)

is strongly inhomogeneous in the state (A) unless a or b is zero; in fact, its θ-dependence is given by

$$\rho(\theta) \propto |a|^2 + |b|^2 + 2\,\mathrm{Re}\,a^*b\,\cos 2m\theta\ .$$

(2.7)

---

[4]The non-θ-dependent wave function $\chi(\mathbf{r}_\perp)$ plays no role in the argument, so for simplicity we have assumed it to be the same for all particles.

By contrast the expectation value of the quantity (2.6) in the state (B) is quite independent of θ; this can easily be seen from the fact that the quantity (2.6) is a sum of one-particle operators and hence cannot have matrix elements between the two components of the linear superposition (B), which differ in the behavior of a macroscopic number of particles. The upshot is that, to the extent that we confine ourselves to low-lying states of the system, states of type (A) are eliminated by the interaction but those of type (B) are still possible. Of course, in the presence of interactions the individual components will no longer have the simple product form $\prod_i \exp(\pm im\theta_i)$, but we should still be able to identify the lowest states of the interacting system which correspond to angular momentum Nm and $-Nm$ respectively (and, in fact, there should still be a macroscopic average number of particles in the single-particle state $e^{\pm im\theta}$). Since the product form is not qualitatively misleading, I shall continue to use it as a schematic representation of the system wave function. For the moment, let us concentrate on the special case of (2.5) obtained by letting either A or B equal zero, that is, schematically

$$\Psi_{\pm}(\mathbf{r}_1\,\mathbf{r}_2\ldots\mathbf{r}_N) = \prod_{i=1}^{N} \chi(\mathbf{r}_{i\perp})\exp(\pm im\theta_i) \quad . \tag{2.8}$$

The next step is to incorporate a Josephson junction in the ring (or rather the analog for neutral bosons, which would be a region of potential sufficiently high that a single particle could not pass classically but must tunnel through the region). The single-particle wave functions are now modified and are no longer eigenfunctions of angular momentum (since the junction destroys the rotational invariance). In fact, we expect that within the bulk ring we still have

$$\phi_{\pm}(\mathbf{r}) \sim \chi(\mathbf{r}_{\perp})\exp(\pm im'\theta) \quad , \tag{2.9}$$

but there is now a finite discontinuity of phase $\Delta\phi$ across the junction. (For present purposes the detailed behavior of the wave function inside the junction is unimportant — though cf. below.) As a result the "quantization" condition for m (or now m'), which results from the requirement of single-valuedness of the wave function, now becomes[5]

$$m' \pm \frac{\Delta\phi}{2\pi} = m \quad , \tag{2.10}$$

where m is an integer as before. The value of $\Delta\phi$ is obtained by minimizing the total energy, which is a sum of the bulk kinetic energy (which will be

[5]We are implicitly assuming here that the thickness of the junction is negligible compared to the circumference of the ring. The generalization is obvious.

proportional to m'^2) and the energy associated with the junction, i.e. the energy necessary to distort the single-particle wave function across the junction so as to produce a phase jump of $\Delta\phi$ between the two sides. Evidently this energy must be a periodic function of $\Delta\phi$ with period 2π, but in general it may be multiple-valued. In the simplest cases, however, the wave function inside the junction cannot tolerate a distortion of more than $\pm\pi$ without very appreciable increase in energy, and in such cases the junction energy is a single-valued function of $\Delta\phi$, periodic with period 2π. The simplest such function (which is often realized in the case of actual physical interest, see below) is simply const. $(\cos\Delta\phi)$.

The total wave function of our N-boson system corresponding to (2.8) in the case of a ring with junction will then be, schematically, of the form

$$\Psi_{\pm}(\mathbf{r}_1 \ldots \mathbf{r}_N) = \prod_{i=1}^{N} \phi_{\pm}(\mathbf{r}_i) \quad , \tag{2.11}$$

where $\phi_{\pm}(\mathbf{r})$ is the single-particle wave function described above. At this stage we come across a problem. Are the single-particle wave functions $\phi_{\pm}(\mathbf{r}_i)$ actually energy eigenfunctions as we assumed above? In fact they are not. Because they are related to one another by time reversal, we can always choose instead the linear combinations

$$\phi'(\mathbf{r}) \equiv \frac{1}{\sqrt{2}}(\phi_{+}(\mathbf{r}) \pm \phi_{-}(\mathbf{r})) \quad , \tag{2.12}$$

and a little calculation shows that, in contradistinction to the case of an uninterrupted ring, the two states (2.12) will be somewhat different in energy because of the effect of the junction potential. Consequently, the functions $\phi_{\pm}(\mathbf{r}_i)$ cannot in general be energy eigenfunctions, and it is at first sight tempting to substitute the functions (2.12) for those in Eq. (2.11). However, this would be quite wrong. The resultant function would have strong variations in the particle density as a function of θ, and as argued above would therefore have a large repulsive energy. The (simplest) correct many-body states are indeed approximately of the form (2.11).

We bring our fictitious boson problem as near as possible to the situation of real experimental interest by adding the last complication, namely the interaction with the electromagnetic field. This gives rise, apart from the microscopic effects which can be added into the nonelectromagnetic repulsion, to the usual Coulomb forces between static charge densities and Ampère interactions between the current and the magnetic flux. If the system is superfluid (as would be the case for the Bose-condensed system of charged bosons, see e.g., Schafroth (1955), then the latter interactions have a drastic qualitative effect: currents are induced in the surface of the bulk ring in such a way as to screen the magnetic field out entirely from the interior of the bulk (the Meissner effect). In an uninterrupted bulk ring this leads to the well-known result of flux quantization. The total flux trapped through the ring (i.e. the externally applied flux plus the generated by the circulating currents) is an integral multiple of h/e (for bosons). In a ring with a junction things are a little different. Because the electric current, which for a single particle is proportional to the operator $(-i\hbar\nabla - e\mathbf{A})$, must vanish far inside the bulk ring (because of the Meissner effect) we have $m' = \Phi/(h/e)$ and hence the "single-valuedness" condition on the phase jump across the junction now becomes

$$\Delta\phi = 2\pi\Phi/(h/e) + 2n\pi \quad , \tag{2.13}$$

where Φ is the total flux trapped through the ring.[6] As before, the actual value of $\Delta\phi$ must be obtained by minimizing the total energy. However, while the contribution of the junction is (at least in the simplest cases) unchanged by the gross long-range electromagnetic effects, the kinetic energy associated with the bulk wave function is, for reasonable values of the parameters, now totally dwarfed by the self-inductance energy of the circulating currents. Since this is equal to $\frac{1}{2}(LI^2)$, where L is the self-inductance and I the circulating current, and the total flux is given by the expression $\Phi_x + LI$, where Φ_x is the externally applied component, we have for the self-inductance energy

$$E_{S.I.} = \frac{1}{2}(\Phi - \Phi_x)^2/L \quad , \tag{2.14}$$

while, from (2.13), the junction energy can be written as (const.)$\cos[2\pi\Phi/(h/e)]$. Note that the total energy may or may not have metastable states as a function of Φ, depending on the values of the parameters (cf. below).

Finally we are in a position to describe the real situation, that of the electrons in a superconducting ring interrupted by a Josephson junction. The difference from the boson case is that the electrons, being fermions, form Cooper *pairs*, which however in (or near) the thermodynamic equilibrium state are automatically Bose-condensed; thus the wave function of a superconductor is, schematically, of the form

$$\Psi(\mathbf{r}_1 \mathbf{r}_2 \ldots \mathbf{r}_N) = A\phi(\mathbf{r}_1 \mathbf{r}_2)\phi(\mathbf{r}_3 \mathbf{r}_4) \ldots \phi(\mathbf{r}_{N-1} \mathbf{r}_N) \quad , \tag{2.15}$$

where for notational simplicity we neglect the spins and assume the number of electrons N to be even, and where A represents the operation of total anti

---

[6]For details of this argument (in the superconducting case) see e.g., Bloch (1970).

symmetrization. Were it not for this antisymmetrization, the wave function (2.15) would simply correspond to a Bose condensation of diatomic molecules into the two-particle state $\phi(\mathbf{r}_1 \mathbf{r}_2)$. Guided by this analogy, let us separate the pair wave function[7] $\phi(\mathbf{r}_1 \mathbf{r}_2)$ into center-of-mass and relative coordinates $\mathbf{R}, \rho$. Then the dependence on ρ is irrelevant in the present context, and all the statements we made above about the single-particle wave functions $\phi(\mathbf{r})$ for the Bose system can be transcribed[8] into statements about the electron pair wave function $\phi(\mathbf{R}, \rho)$ regarded as a function of the center-of-mass variable $\mathbf{R}$. The only difference is that the pair has a charge $2e$ rather than e, so that the quantization condition on the flux in an uninterrupted ring would now be $\Phi = n\Phi_0$, where $\Phi_0 \equiv h/2e \cong 2 \times 10^{-15}$ Wb is the "flux quantum" as normally defined in the context of superconductivity theory. Thus the total energy of the SQUID ring as a function of the total flux Φ trapped through it has, within the framework of the model, the form

$$U(\Phi) = \frac{(\Phi - \Phi_x)^2}{2L} - \frac{I_c \Phi_0}{2\pi} \cos(2\pi\Phi/\Phi_0) \quad , \qquad (2.16)$$

where we have arbitrarily (at first sight!) defined the constant in the expression for the junction energy as $I_c \Phi_0/2\pi$, where I_c has the dimensions of current.

Equation (2.16) is very familiar in the literature on superconducting devices, and can in fact be derived very simply from basic phenomenological considerations concerning the Josephson effect (Lounasmaa, 1974); I have chosen to give a more long-winded "derivation" here to emphasize the connection with the actual many-body wave function of the system, which will play a crucial role later on. A number of points about Eq. (2.16) should be noted. First, it gives the energy of the ring quite generally as a function of the trapped flux Φ; the latter is assumed[9] to be related by (2.13) (with $h/e \approx h/2e$) to the phase jump $\Delta\phi$ of the center-of-mass wave function of the Cooper pairs across the

---

[7] This is not the quantity $F(\mathbf{r}_1 \mathbf{r}_2) \equiv \langle \psi_\uparrow^+(\mathbf{r}_1)\psi_\downarrow^+(\mathbf{r}_2) \rangle$ which is often called the "pair wave function" in standard accounts of superconductivity theory: the quantities are related by a rather unpleasant integral equation, see e.g., Ambegaokar (1969).

[8] The argument leading to the conclusion that states of the form (2.4) are excluded at low energy is slightly more subtle in the superconducting case, since (in view of the strong effects of antisymmetrization) there is no direct relation between the magnitude of the pair wave function (in either sense of the words) and the total electron density. However, the conclusion (which can be obtained, e.g., from a Landau-Ginzburg analysis) is unchanged.

[9] This relation should hold provided that the phenomena of interest occur on a frequency scale small compared to the bulk plasma frequency ($\sim 10^{16}$ s^{-1}). This is almost always true in the applications we shall consider below.

the junction. Thus, in effect the energy is a function of a quantity, $\Delta\phi$, which characterizes a quantum-mechanical wave function. Equation (2.16) applies whether or not the equilibrium value of $\Delta\phi$ (or equivalently Φ) is achieved. The condition for equilibrium is given by minimizing it, that is,

$$\frac{\Phi - \Phi_x}{L} = I_c \sin(2\pi\Phi/\Phi_0) \quad , \qquad (2.17)$$

or, since the circulating current I in the ring is related to Φ by $\Phi = \Phi_x + LI$, and we have $2\pi\Phi/\Phi_0 = \Delta\phi + 2\pi n$,

$$I = I_c \sin\Delta\phi \quad , \qquad (2.18)$$

which is just the usual expression for the current through a Josephson junction as a function of the phase difference across it. It should be carefully noted that if the quantity $2\pi L I_c/\Phi_0$ (which is often denoted β_L in the literature) is greater than unity, Eq. (2.17) in general will have more than one solution, corresponding in general to metastable minima of the energy as well as to the stable one (which is given by the solution of Eq. (2.17) for the Φ nearest to Φ_x). A case which will be of special interest for our purposes is when Φ_x is exactly $\frac{1}{2}\Phi_0$ and $\beta_L > 1$. In this case it may be verified that there are two degenerate absolute minima of Eq. (2.17), symmetrically placed with respect to the value $\frac{1}{2}\Phi_0$.

A second point to note regarding Eq. (2.16) is the order of magnitude of the energies involved, in particular, of the second term which determines the maximum height of the energy-barriers between the various metastable or stable wells. Since the numerical value of the flux quantum Φ_0 is about 2×10^{-15} Wb, and typical values of critical currents in most Josephson junctions are in the range $1\,\text{nA} - 1\,\text{mA}$, the value of the energy barrier is, crudely, in the range $10^{-24} - 10^{-18}$ J. This is clearly by any reasonable standards a *microscopic* quantity (for comparison, the thermal energy of a single atom at room temperature is of order 10^{-20} J).

Despite this, however, the states corresponding to different energy minima are, by any reasonable standards, *macroscopically* different. The difference in the circulating currents is, from Eq. (2.18), in general of the order of the critical current, that is (for realistic experiments — see below), of the order of a few microamperes. Moreover, the common wave function of all the $N/2$ Cooper pairs in the system is different for different values of Φ, since these correspond to different phase jumps across the junction.[10] Thus, we have achieved the first two of the desiderata specified above.

We now come to a crucial point. So far, we have assumed that the quantum-mechanical many-body wave function of the system we are considering corresponds (apart from antisymmetrization, etc.) to Bose condensation of the Cooper pairs into a single two-particle state whose center-of-mass wave function has the general behavior specified, and in particular has a definite jump $\Delta\phi$ in phase across the junction; since we have not relaxed the condition $\Delta\phi = 2\pi(\Phi/\Phi_0 + n)$ which follows from the Meissner effect (and will not do so below) specification of $\Delta\phi$ is equivalent to specification of the trapped flux Φ. Thus the many-body wave functions corresponding to the states considered so far are of the general form

$$\Psi(\mathbf{r}_1 \mathbf{r}_2 \ldots \mathbf{r}_N) = A\,\chi(\mathbf{r}_1\mathbf{r}_2:\Phi)\chi(\mathbf{r}_3\mathbf{r}_4:\Phi)\ldots\chi(\mathbf{r}_{N-1}\mathbf{r}_N:\Phi)$$

$$\equiv \Psi_\Phi(\mathbf{r}_1\mathbf{r}_2\ldots\mathbf{r}_N) \equiv |\Phi\rangle \quad . \tag{2.19}$$

Now, in our original model of noninteracting bosons these would correspond to the simple product states of the form

$$\Psi(\mathbf{r}_1\mathbf{r}_2\ldots\mathbf{r}_N) = \prod_i \exp(im\theta_i) \equiv \Psi_m(\mathbf{r}_1\mathbf{r}_2\ldots\mathbf{r}_N) \tag{2.20}$$

(where m is the same for all particles). However, in that case we saw that we can also write down states of the form

$$\Psi(\mathbf{r}_1\mathbf{r}_2\ldots\mathbf{r}_N) = \sum_m a_m \Psi_m(\mathbf{r}_1\mathbf{r}_2\ldots\mathbf{r}_N) \,, \tag{2.21}$$

and even that for certain choices (e.g., a_m nonzero for $m = m_0$ and $-m_0$ only, cf. Eq. (2.5)) these can actually be energy eigenfunctions. Moreover, we saw that, in contradistinction to states of the form (2.4), states of the type (2.19) are not suppressed by the introduction of repulsive interparticle interactions. There seems no obvious reason why the further complications added to the model (insertion of a junction, coupling to the electromagnetic fields, transition from bosons to Cooper pairs) should change this qualitative result. Consequently,

---

[10] Moreover, the degree of orthogonality of the pair wave functions corresponding to appreciably different values of Φ is of order 1. At first sight this is slightly surprising since the bulk of the system (i.e. the inside of the thick superconducting ring) is apparently in "the same" state (carrying no current) whatever the value of Φ. However, the θ-dependent quantum mechanical phase factor is different for different Φ, leading to orthogonality. Whether or not this particular observation is actually relevant for the foundations of quantum mechanics is a delicate issue: compare the discussion of a related point in Leggett (1980), Sec. 3.

we are led to consider the possibility of quantum-mechanical states which correspond to *superposition* of states of the type (2.19) with different values of Φ, i.e.

$$\Psi(\mathbf{r}_1 \mathbf{r}_2 \ldots \mathbf{r}_N) = \sum_m a_m \Psi_{\Phi_m}(\mathbf{r}_1 \mathbf{r}_2 \ldots \mathbf{r}_N) \equiv \sum_m a_m |\Phi_m\rangle . \quad (2.22)$$

However, since the flux is a continuous variable, we should actually generalize this to read

$$\Psi(\mathbf{r}_1 \mathbf{r}_2 \ldots \mathbf{r}_N) = \int a(\Phi) \Psi_\Phi(\mathbf{r}_1 \mathbf{r}_2 \ldots \mathbf{r}_N) d\Phi \equiv \int a(\Phi)|\Phi\rangle d\Phi ,$$

$$(2.23)$$

and then write

$$a(\Phi) \equiv \Psi(\Phi) , \quad (2.24)$$

so that the many-body wave function is now regarded as *a function of the variable* Φ. (This last step tends to cause some confusion, but it is actually no different in principle from what we implicitly do in setting up the quantum mechanics of a simple with coordinate x: we first define wave functions $|x\rangle$ which correspond to the particle being definitely at point x, then define a general quantum-mechanical state by giving its amplitude in this basis, i.e.,

$$\Psi = \int a(x)|x\rangle dx \quad (2.25)$$

where the amplitude $a(x)$ is just what we normally write as the wave function $\psi(x)$. Cf. Dirac (1958)). Note that if the wave function $\Psi(\Phi)$ has appreciable amplitude for substantially different values of Φ, the state (2.23) is *a linear superposition of macroscopically different states.*

Are states of the form (2.23) of any practical interest? If the (relevant part of the) Hamiltonian were indeed given completely by Eq. (2.16), the answer would be no. Obviously, the energy eigenfunctions would then be given by the "flux eigenfunctions" (Eq. 2.17), and even in the event of superposition of two degenerate eigenstates there would be no observable effects; the situation would be similar to that of a particle with potential energy but no kinetic energy, i.e. infinite mass. In fact even the *classical* dynamics obtained from Eq. (2.16) would be completely trivial. To generate nontrivial dynamics we need to add to the Hamiltonian a term corresponding to a "kinetic energy."

A clue as to the possible nature of such a term is obtained by noticing that according to Maxwell's equations, if we assume that the time variation of quantities is slow enough that no electric fields arise in the bulk superconductor, the time derivative of the flux trapped through the ring is equal to the voltage developed across the junction; so that if there is any capacitance associated with the junction, we should expect there to be associated with this an extra energy K of the form

$$K = \tfrac{1}{2}CV^2 = \tfrac{1}{2}C\dot{\Phi}^2 \ . \tag{2.26}$$

If, now, we could treat Φ as a classical variable and incorporate it in a Lagrangian $L \equiv K - U$ (where $U(\Phi)$ is the quantity given in Eq. (2.16)) we would find that the momentum conjugate to Φ is

$$p_\Phi = \partial L/\partial \dot{\Phi} = C\dot{\Phi} = CV = Q \ , \tag{2.27}$$

where Q is the charge transferred across the junction.[11] If, we further assume that we can quantize the system according to the standard prescriptions for a classical system, we find that Q becomes an operator:

$$Q = p_\Phi \rightarrow -i\hbar \partial/\partial \Phi \ , \tag{2.28}$$

and the "kinetic energy" $K = Q^2/2C$ becomes the operator $(-\hbar^2/2C)(\partial^2/\partial\Phi^2)$. Thus we obtain for our system a Schrödinger equation and Hamiltonian H given by

$$i\hbar \frac{\partial \Psi(\Phi)}{\partial t} = \hat{H}\Psi(\Phi), \qquad \hat{H} = \frac{-\hbar^2}{2C} \frac{\partial^2}{\partial\Phi^2} + U(\Phi) \tag{2.29}$$

where $U(\Phi)$ is the quantity given in Eq. (2.16). It is clear that, if Φ is regarded as analogous to a coordinate x of a particle, then the capacitance C of the junction is exactly analogous to the mass of the particle.

Is the replacement (2.28) in fact legitimate? There are a number of possible ways of justifying it, none perhaps completely rigorous but all persuasive. One is simply to note that only if we make this replacement does the Hamiltonian in Eq. (2.29) lead, in the semiclassical limit, to the classical dynamics obtained by simple macroscopic circuit analysis (cf. below). A second way (Leggett, 1984b) is to start from the standard quantum electrodynamic commutation relations (Gasiorowicz, 1966) of the electric and magnetic fields and make an appropriate coordinate transformation. A third argument, which is perhaps the most natural one to use in conjunction with the present development of the subject, is analogous to the one used originally by Anderson for a simple

---

[11] The sign associated with V and Q is a matter of convention and clearly does not effect the subsequent argument.

Josephson junction (Anderson, 1964) and elaborated by the author (Leggett, 1966) for a two-band superconductor. For simplicity I give it here explicitly for the case of charged bosons (the fermion nature of the real system introduces some notational complications (cf. Leggett (1966)), but the eventual outcome is the same). Consider a product many-body wave function of the (interacting and charged) boson system in the ring closed with a junction, which corresponds to a superposition of different flux eigenfunctions:

$$\Psi(\mathbf{r}_1 \ldots \mathbf{r}_N) = \int d\Phi \, a(\Phi) \prod_{i=1}^{N} \chi(\mathbf{r}_i : \Phi) \quad , \tag{2.30}$$

$$a(\Phi) \equiv \Psi(\Phi) \ . \tag{2.31}$$

The operator $\partial/\partial\Phi$, when acting on the many-body wave function $\Psi(\mathbf{r}_1 \, \mathbf{r}_2 \ldots \mathbf{r}_N)$, must be interpreted in the following way:

$$\frac{\partial}{\partial\Phi} \Psi(\mathbf{r}_1 \, \mathbf{r}_2 \ldots \mathbf{r}_N) \equiv \int d\Phi \, \frac{da(\Phi)}{d\Phi} \prod_{i=1}^{N} \chi(\mathbf{r}_i : \Phi) \ . \tag{2.32}$$

Note that the quantity $\prod_{i=1}^{N} \chi(\mathbf{r}_i : \Phi) \equiv |\Phi\rangle$ corresponds to the basis vector in the flux representation and is therefore not differentiated.[12] If now we integrate the right-hand side of Eq. (2.32) by parts and assume that $a(\Phi)$ vanishes for sufficiently large positive of negative Φ (as must be the case for all reasonable states) we obtain

$$-i\hbar \, \frac{\partial}{\partial\Phi} \Psi(\mathbf{r}_1 \, \mathbf{r}_2 \ldots \mathbf{r}_N)$$

$$= +i\hbar \int d\Phi \, a(\Phi) \sum_{j=1}^{N} \frac{\partial}{\partial\Phi} \ln\chi(\mathbf{r}_j : \Phi) \prod_{i=1}^{N} \chi(\mathbf{r}_i : \Phi)$$

$$\equiv \left(+i\hbar \sum_{j=1}^{N} \frac{\partial}{\partial\Phi} \ln\chi(\mathbf{r}_j : \Phi) \right) \Psi(\mathbf{r}_1 \, \mathbf{r}_2 \ldots \mathbf{r}_N) \ . \tag{2.33}$$

When Eq. (2.33) is multiplied by a second many-body wave function $\Psi'(\mathbf{r}_1 \, \mathbf{r}_2 \ldots \mathbf{r}_N)$ and integrated over its arguments, the overwhelmingly dominant contribution will come from regions where all the $\mathbf{r}_j$ are in the bulk ring.

---

[12] Readers who find this point difficult are recommended to carry out explicitly the corresponding procedure for the momentum operator $-i\hbar\partial/\partial x$ for a simple one-dimensional system using the Dirac notation.

Thus we can take the single-particle wave functions to have the appropriate "bulk" form, i.e. (neglecting the dependence on the coordinates in the cross-section $r_{\perp j}$)

$$\chi(r_j : \Phi) = \text{const. } e^{im'\theta_j} , \quad m' \equiv \Phi/(h/e) \quad . \tag{2.34}$$

It is convenient to choose the origin of θ so that the position of the junction corresponds to $\theta = \pi$. Then we see from Eqs. (2.33) and (2.34) that

$$-i\hbar \frac{\partial}{\partial \Phi} \Psi(r_1 r_2 \ldots r_N) = -\sum_{j=1}^{N} \left(\frac{e\theta_j}{2\pi} \right) \Psi(r_1 r_2 \ldots r_N) \quad . \tag{2.35}$$

The operator $\hat{\Omega} \equiv \sum_{j=1}^{N} e\theta_j/2\pi$ is clearly a measure of the departure of the charge displacement from a uniform state. In a realistic situation, the Coulomb forces within the bulk ring will tend to localize the extra charge close to the edges of the junction;[13] since, when a single charge e is transferred across the junction, $(\theta_j = \pi \rightarrow \theta_j = -\pi)$ the quantity $\hat{\Omega}$ changes by e, we can say that as regards the low-energy states of the system, $\hat{\Omega}$ is equivalent to the operator of the charge localized on the capacitance associated with the junction. Thus Eqs. (2.28) and (2.29) are justified.[14]

The argument goes through similarly, but with some notational complications due to the necessity of antisymmetrization, when the condensed objects are Cooper pairs rather than bosons: the condition (2.34) on m' is now replaced by $m' = \Phi/(h/2e) (\equiv \Phi/\Phi_0)$, but the extra factor of 2 in Eq. (2.35) is cancelled by the fact that θ_j now represents the (center-of-mass) position of a pair rather than a single particle.

The above argument may no doubt sound on first reading less than totally convincing. The plausibility of the conclusion is however strengthened by noting that it does at least give "common-sense" classical dynamics. To see this, we note that in the classical limit we can replace the operator $-\hbar^2 \partial^2/\partial \Phi^2$ in the Hamiltonian in Eq. (2.29) by the classical quantity p_ϕ^2, where as above

---

[13] This is true provided that the capacitance of the junction (and its immediate surroundings) is much larger than that of the bulk ring — a caveat which should be borne in mind when considering the limit of ultra-small junction capacitance.

[14] There are, of course, a number of points in this argument which need to be tidied up. In particular one might wonder whether the contribution to Eq. (2.33) from the junction itself might give difficulties when the single-particle wave function $\chi(r_j:\Phi)$ has a node there (as is liable to happen for $\Delta\phi = \pi$). In fact *prima facie* it does not, as may be seen by a detailed consideration of the behavior of the χ's within the junction; however, it is possible there on further subtleties to be explored here.

$p_\phi = Q = CV = C\dot{\Phi}$. The resultant classical equation of motion is then

$$C\ddot{\Phi} = -\partial U(\Phi)/\partial \Phi \quad , \tag{2.36}$$

or, using the relation $\Phi = \Phi_x + LI$,

$$\dot{Q} = C\ddot{\Phi} = I - I_c \sin 2\pi\Phi/\Phi_0 \tag{2.37}$$

which is just the equation of conservation of charge flowing into the junction region. (I is the total current flowing in, $I_c \sin 2\pi\Phi/\Phi_0 = I_c \sin \Delta\phi$ is the Josephson current and the difference is stored on the junction capacitance: cf. Fig. 4, ignoring the resistance.)

In any case, it should be strongly emphasized that the Hamiltonian in Eq. (2.29) can only be expected to give a reasonable description of the quantum-mechanical behavior of the system under very restricted conditions. In particular, the characteristic frequencies ω which occur in the resultant motion certainly must be less than the plasma frequency ω_p ($\sim 10^{16}$ s^{-1}), otherwise the Meissner effect and hence the relation between Φ and $\Delta\phi$ breaks down. For a SQUID ring of reasonably "macroscopic" geometrical size (say ~ 1 mm $- 1$ cm diameter) there is actually a much more stringent condition. Clearly, if during a cycle of oscillation there is no time for a light wave to propagate across the ring, then the whole idea of describing its flux state in terms of a single macroscopic variable Φ breaks down and we need to think of it as something more like a transmission line, with appropriate distributed capacitances and inductances. This restricts the simple model in Eq. (2.29) to frequencies $\omega \lesssim 2\pi c/l$, where l is a characteristic dimension of the ring; e.g., for $l = 1$ mm we need $\omega \lesssim 2 \times 10^{12}$ s^{-1}. In our subsequent analysis we shall be dealing only with situations which satisfy these conditions by an adequate margin.

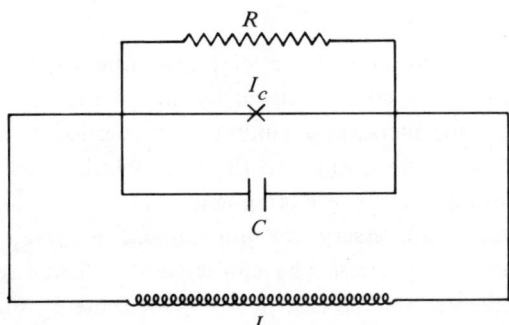

Fig. 4. Equivalent circuit for the "resistively shunted junction" model of a SQUID ring.

Given that we believe that Eq. (2.29) is an adequate zeroth-order approximation to the true Hamiltonian of our system as regards the quantum mechanics of the flux, we can proceed to calculate various transition probabilities, etc., just as we would for a particle of mass m, with coordinate x, moving in a potential $V(x)$. We simply make everywhere the replacements $x \to \Phi$, $V(x) \to U(\Phi)$, $m \to C$. In particular we can calculate (a) the zero-point uncertainty $\Delta\Phi_{\text{rms}}$ in the value of the flux when the system is in a stable or metastable potential minimum, (b) the rate Γ of "macroscopic quantum tunneling" (MQT) out of a metastable potential well and (c) the rate Δ of oscillation between two degenerate potential minima ("macroscopic quantum coherence" or MQC). The calculations are straightforward provided that we are in the WKB limit; this turns out to be roughly equivalent to the condition $\lambda \equiv (8CI_c\phi_0^3/\pi^3\hbar^2)^{1/2} \gg 1$, and is well satisfied for $CI_c \gtrsim 2 \times 10^{-23}$ farad amps, a condition which, so far at least, has been met in most if not all real-life experiments. The details of the calculations in this limit are given in Leggett (1984b) and I quote here only the results:

$$\Delta\Phi_{\text{rms}} = \pi^{-1}\lambda^{-1/2}\,\Phi_0 \quad ; \tag{2.38}$$

$$\Gamma = \frac{144}{5\pi}\left(\frac{2\pi I_c}{C\Phi_0}\right)^{1/2} \cdot \lambda \cdot \left(\frac{\delta\Phi_x}{\Phi_0}\right)^{3/2}$$

$$\times \exp-\left\{\frac{24}{5}\cdot 2^{-1/4}\lambda\left(\frac{\delta\Phi_x}{\Phi_0}\right)^{5/4}\right\} \quad ; \tag{2.39}$$

$$\Delta = \text{const.}\,(\beta_L - 1)^{1/2}\left(\frac{2\pi I_c}{C\Phi_0}\right)^{1/2} \exp\{-2^{-1/2}(\beta_L-1)^{3/2}\lambda\} \quad . \tag{2.40}$$

Equation (2.39) refers to the realistic MQT case, where $\beta_L \equiv 2\pi L I_c/\phi_0$ is large compared to 1 and the system is biased by an external flux $\delta\Phi_x$ below that which would cause the metastable minimum in question to become unstable. Equation (2.40), by contrast, refers to the case (which is the realistic one for a macroscopic quantum coherence experiment) where β_L is close to unity. It is straightforward, but unnecessary for our present purpose, to calculate the corrections for larger β_L^{-1} in case (b) and larger $\beta_L - 1$ in case (c). The result (a) is quoted explicitly for the case $\beta_L \to \infty$; for finite β_L the result is simply multiplied by $(1 + \beta_L)^{-1/2}$.

Before concluding this section, let me say a word about the transcription of the above results to current-biased junctions. It is tempting to regard a junction biased by a fixed external current I_{ext} as simply the limit of the corresponding SQUID ring with $L \to \infty$ and $\Phi_x/L \to I_{ext}$, with the variable Φ replaced by $(\Phi_0/2\pi)\Delta\phi$. In that case the appropriate Hamiltonian would be

$$\hat{H} = -\frac{\hbar^2}{2C}\left(\frac{2\pi}{\phi_0}\right)^2 \frac{\partial^2}{\partial(\Delta\phi)^2} - \frac{\Phi_0}{2\pi}[I_{ext}\Delta\phi + I_c\cos\Delta\phi] \quad .(2.41)$$

The corresponding *classical* Hamiltonian (i.e. with the first term replaced by $\frac{1}{2}C(\phi_0/2\pi)^2\dot{\phi}^2$) is indeed routinely used to describe the classical behavior of a current-biased junction, with results in reasonable agreement with experiment (though cf. next section). However, the use of Eq. (2.41) for a quantum-mechanical description raises some very delicate questions, associated with the facts that (a) the "minima" of the potential energy in Eq. (2.41) (often known in the literature as the "washboard potential"), which appear to occur at values of $\Delta\phi$ separated by 2π, are *prima facie* actually not physically distinct, and (b) the concept of a classical "torque" (here I_{ext}) applied to a periodic quantum-mechanical variable is by no means a clear one. I shall sidestep these questions here[15] by assuming that as long as the quantum-mechanical motion of interest to use involves values of $\Delta\phi$ must less than 2π, the periodicity has no effect and Eq. (2.41) is a valid description, and will later consider only phenomena which occur under these conditions. The assumption that under such circumstances there is no qualitative difference between the behavior of a current-biased junction and that of a SQUID ring is buttressed by (a) the results of Eckern *et al.* (Ambegaokar *et al.*, 1982; Eckern *et al.*, 1984) (Sec. 4) and (b) the similarity of the experimental results on MQT obtained in the two systems (cf. end of Sec. 4).

3. The Effects of Coupling to the "Environment"

In the previous section we "derived" a Schrödinger equation (Eq. (2.29)) for the flux trapped in the SQUID ring. The reader could be forgiven if he or she finds the "derivation" less than totally satisfactory. Indeed, quite apart from the various points of detail which have been less than adequately discussed in the argument, there are two rather strong *a priori* reasons (which I shall argue below are not unconnected) why one might view the result in Eq. (2.29) with skepticism. The first goes back to the general observation, so often made in formal discussion of the quantum theory of measurement, which was discussed

---

[15] If one wants to consider phenomena of the type discussed in Widom *et al.* (1981) and Rogovin and Nagel (1982), it is essential to confront these problems head on. Cf. Likharev and Zorin (1985). This quantum has recently been considerably clarified by W. Zwerger, A. T. Dorsey and M. P. A. Fisher (preprint).

in Sec. 1: no macroscopic variable is ever sufficiently decoupled from its environment that it can be described by a wave function in its own right, rather the wave function of the "universe" must be of the form (1.8) (or an appropriate generalization) which *inter alia* will automatically make quantum-mechanical effects in the motion of the macroscopic variable unobservable. In particular the states which would be predicted by Eq. (2.29) to occur in an "MQC" (ammonia-inversion-resonance) type situation, namely, schematically,

$$\Psi(\Phi) = a(t)\Psi_R(\Phi) + b(t)\Psi_L(\Phi) \quad , \tag{3.1}$$

where Ψ_R and Ψ_L correspond to states localized in the "right" and "left" wells (i.e. corresponding to clockwise and counterclockwise circulating currents respectively) might be thought a totally unrealistic description of the physical situation. In practice the "environment" of the flux in a SQUID ring might include the normal component, the radiation field, the electrons in any shunt conductances, the phonons of the metal and much else besides, and it seems *prima facie* inconceivable that in so complex an environment the wave function could remain even approximately of the form (3.1) rather than (1.8).

The second *a priori* objection is that the remark which was used to bolster the argument of the previous section, namely that Eq. (2.29) at least reduces to the intuitively correct classical dynamics when we take the semiclassical limit, is actually not correct. In fact, the Eq. (2.36) which results from taking the classical limit of Eq. (2.29) is never obeyed by any SQUID ring in real life. At the very least, to obtain an even approximately realistic description of the experimentally observed behavior we must incorporate a phenomenological conductance $1/R$ in parallel with the junction, so that the circuit diagram looks like that shown in Fig. 4. The equation of motion which then follows from a simple circuit analysis is

$$C\ddot{\Phi} + \dot{\Phi}/R + \partial U(\Phi)/\partial \Phi = 0 \quad , \tag{3.2}$$

with $U(\Phi)$ given by Eq. (2.16). For a current-biased junction the analogous equation for the phase difference $\Delta\phi$ is

$$C\Delta\ddot{\phi} + \frac{\Delta\dot{\phi}}{R} + \frac{2\pi I_c}{\phi_0}\sin\Delta\phi = \frac{2\pi I_{ext}}{\phi_0} \quad . \tag{3.3}$$

The classical model (McCumber, 1968; Stewart, 1968) which leads to these equations is known as the "resistively shunted junction" model and has been widely used in the literature to interpret the experimental behavior of tunnel junctions. However, not all junctions show behavior which can be explained in terms of Eq. (3.3). In particular, the idea of a linear resistance which is completely independent of frequency seems inapplicable in many cases. From a theoretical point of view it is easy to understand this. The characteristic frequency which emerges from Eq. (3.3) is typically of the order of the Josephson plasma frequency

$$\omega_J \equiv (2\pi I_c / C\phi_0)^{1/2} \quad,$$

(or lower). On the other hand, we would expect from our microscopic model of the superconducting state that the response of the system changes markedly (even at zero temperature) as soon as ω becomes greater than twice the single-particle energy gap Δ, since at that point real excitation (breakup) of Cooper pairs becomes possible. If, therefore, ω_J is comparable to 2Δ (as it may well be in many real-life cases) we should certainly not expect the RSJ model to apply in unmodified form over the whole frequency range of interest. On the other hand, at frequencies much lower than 2Δ it is not unreasonable to expect it to apply, at least to a first approximation. In this case, at the low temperatures which are relevant to the experiments to macroscopic quantum mechanics, the shunting conductance $1/R$ certainly cannot be due to the normal component (there is none[16]) and is usually ascribed to metallic shorts on the junction surface or other effects whose detailed origin is uncertain. However, whatever the detailed features or uncertainties of the model, one thing is certain: in real life both current-biased junctions and SQUIDs show considerable effects of dissipation. At first sight this totally spoils our supporting argument for Eq. (2.29).

The point of view which is implicit in the work of the present author and his collaborators, and of some other groups working in this area, has been that the above two objections can to a large extent be played off against one another, in the following sense. Let us agree that in real life it is hopelessly optimistic to try to describe the behavior of a macroscopic variable such as the flux Φ in terms of wave function $\Psi(\Phi)$ which ignores the effect of the environment, and that we should rather write the wave function of the "universe" in the form $\Psi = \Psi(\Phi \cdot \xi_1 \xi_2 \ldots \xi_n)$ where the ξ_i are the myriad microscopic co-ordinates characterizing the environment. The time evolution of Ψ should be determine by a Hamiltonian which, *inter alia*, couples the system to the environment:

$$\hat{H} = \hat{H}_s(\Phi, p_\Phi) + \hat{H}_{env}(\{\xi_i, \pi_i\}) + \hat{H}_{int}(\Phi, p_\Phi : \{\xi_i, \pi_i\}) \,, \qquad (3.4)$$

---

[16] For a 1 cm^3 bulk sample of niobium (transition temperature $\sim 9\,$K) at a temperature of 100 mK, it is easily verified that the mean number (not fraction) of normal quasiparticles present in equilibrium is of order 10^{-40} !

where p_Φ and π_i are the momenta conjugate to Φ and ξ_i respectively. Now, it is the total Hamiltonian in Eq. (3.4), including the coupling between system and environment, which will determine the characteristic macroscopic quantum-mechanical behavior of the system (MQT, MQC, etc.), if indeed any occurs. However, this same Hamiltonian in Eq. (3.4) should also determine the quantum-mechanical behavior in the semiclassical limit, i.e. the "classical" equations of motion of the flux, etc., including the dissipative terms. The hope around which much of the work in this area has been built is that *the behavior of the system in the "classical" regime will determine the form of the Hamiltonian to a sufficient degree that we can make nontrivial predictions about the macroscopic quantum behavior.* If we take this point of view, then many of the uncertainties of detail, both in the argument leading to Eq. (2.29) and in the mechanisms of system-environment interaction, are automatically made irrelevant. A complete knowledge of the classical dynamics will (we hope!) determine the macroscopic quantum behavior irrespective of the unknown details of the model.

Before discussing this proposition in detail, we shall note that there is an alternative point of view on this question, which is embodied in particular in the work of Ambegaokar and co-workers (Ambegaokar *et al.*, 1982; Eckern *et al.*, 1984). This is that, at least in certain cases, we have detailed microscopic models of the systems which we are considering which have been sufficiently well confirmed by the good agreement of their predictions with experiment in the context of "semiclassical" phenomena that it is sensible to use them without further ado to make predictions of the behavior of these systems in MQT and MQC experiments. In particular, it is argued that the semiclassical behavior (e.g., the current-voltage characteristics) of many Josephson junctions of the "tunnel oxide" variety is very well described by the standard BCS Hamiltonian for superconductivity in the bulk metal plus the Bardeen-Josephson (Bardeen, 1961; Josephson, 1965) tunneling Hamiltonian for the junction itself. Thus Eckern *et al.* (Ambegaokar *et al.*, 1982; Eckern *et al.*, 1984) apply this Hamiltonian to calculate the functional integrals which, as we shall see below, determine the MQT and MQC behavior.

The point of view of the present author is that both these approaches should be pursued in parallel, but that the relative effort one puts into one or the other will probably be a function of one's expectation and/or hopes concerning the final outcome of the whole program of research and in particular of the outcome, if it can be done, of the MQC experiment. If one has sufficient faith in quantum mechanics as a universal theory of the world to be quite sure that it will describe the result of the experiment correctly, then one's main interest will probably lie in predicting the quantitative details accurately before the event, and then it makes eminent sense to use for this purpose models such as the Bardeen-Josephson model which have been successful elsewhere. If on the hand one has even the slightest suspicion that nature at this level is *not* perfectly described by quantum mechanics, then one will be prepared for a possible outcome to the MQC experiment which is not in agreement with the predictions of one's quantum-mechanical calculations; if one hopes to use any such disagreement as *prima facie* evidence for the breakdown of quantum mechanics at this level, it is clearly essential to pre-empt, as far as possible, the objection that the disagreement was simply a consequence of the fact that the detailed Hamiltonian used for the quantum-mechanical calculations was inadequate. Since no concrete microscopic model, however successful, can ever be rigorously proved to be exact for all purposes, it is clear that in this case it is preferable to base one's arguments, as far as possible, on general considerations which make no reference to such a model. In particular we would like, if possible, to demonstrate (or at least argue plausibly) that *any* Hamiltonian for the system-environment coupling which has the effect of destroying the MQC phenomenon must also necessarily produce predictions for the semiclassical behavior (which everyone — at present at least — agrees should be well described by standard quantum mechanics) which are susceptible to an experimental test. If such semiclassical predictions are not confirmed experimentally, and the MQC phenomenon is nevertheless observed not to occur, we would be in a considerably stronger position to argue for a breakdown of the quantum description in the latter case.

A further reason why one might wish, while recognizing the importance of the results of Ambegaokar and co-workers (Ambegaokar *et al.*, 1982; Eckern *et al.*, 1984), to try to go somewhat beyond them, is that this calculation actually gives a result which at zero temperature (and in fact for any temperature well below the transition temperature of the junction) is completely isomorphic to that obtained from the Hamiltonian of an *isolated* system with a renormalized (and in general phase-dependent, see Sec. 4) capacitance. In real life the dissipation observed at low frequencies ($\sim \omega_J$) in any real junction, e.g. by measuring its current-voltage characteristics,[17] may be very small but it is never strictly zero. This indicates a finite rate of irreversible exchange of energy between the system and its environment, and because it is just this same process which destroys the macroscopic quantum coherence (cf. below) it is important to make quite sure it has been adequately taken into account, however small it

---

[17] The d. c. I-V characteristic actually measures a complicated average of the dissipation over frequency, including some dissipation at frequencies greater than 2Δ. However, it appears unlikely that taking this feature into account will bring the predictions of the Bardeen-Josephson model into agreement with experiment (Chen *et al.*, 1986).

may seem to be. Thus, ideally one should combine the results of the calculation of Eckern *et al.*, and those obtained below (cf. Larkin and Ovchinnikov (1982)).

The above program — of relating the MQC and MQT behavior predicted by a general Hamiltonian for the system-environment coupling to the "classical" behavior predicted by that same Hamiltonian — is rendered more plausible by the recognition of the distinction between "adiabatic" and "irreversible" aspects of the effects of the coupling. For the sake of simplicity let us suppose that the coupling is relatively weak, so that we do not expect it to shift the characteristic (classical) frequency of the system by an order of magnitude, and let this frequency be of order ω_0. Since the variable describing the "system" (e.g., the flux trapped in an rf SQUID ring) is macroscopic by the usual standards, while the many degrees of freedom of the environment generally speaking correspond to microscopic motions, the characteristic frequency scale of the environment, ω_{env}, is typically very much larger than ω_0. (For example, in the case of MQT in a Josephson junction or SQUID ω_0 might be the junction plasma frequency (say $\sim 10^{12} \, s^{-1}$), while (in the case where the normal component is the main contributor to the "environment") ω_{env} would be of the order of the Drude relaxation time, which for a reasonably dirty metal could be say $10^{14} - 10^{15} \, s^{-1}$; thus, $\omega_{env}/\omega_0 \sim 10^2 - 10^3$. In most other systems the ratio is even larger than this.) Now, a point which is central to the general program is that those degrees of freedom of the environment which have frequencies of the order of ω_{env} (hence $\gg \omega_0$) will follow the motion of the system *adiabatically*, and therefore will not automatically destroy the quantum coherence we are looking for. To illustrate this point, let us consider specifically an MQC experiment (which is described in more detail in Sec. 5). In this case, if we were dealing with a totally isolated system, we would describe the motion in terms of the states ψ_L, ψ_R corresponding to being in one or the other of the two degenerate minima of the potential. We would start the system in (say) the left well, and its subsequent state is predicted by quantum mechanics to be a linear superposition of the form (2.21), i.e.

$$\psi(t) = a(t)\psi_L(\Phi) + b(t)\psi_R(\Phi) \quad . \tag{3.5}$$

What we would observe is then the probability $P_L(t)$ for the system to be on the left at a subsequent time t (predicted by quantum mechanics, of course, to be $|a(t)|^2$). Let us now imagine switching on the coupling to the environment. To the extent that this coupling is adiabatic, the correct description of the "universe" is now by a wave function $\Psi(\Phi : \{\xi_i\})$ of the form

$$\Psi(t) = a(t)\psi_L(\Phi)\chi_L\{\xi_i\} + b(t)\psi_R(\Phi)\chi_R\{\xi_i\}$$

$$\equiv a(t)\Psi_L + b(t)\Psi_R \quad , \tag{3.6}$$

— that is, a special case of Eq. (1.8) with $\alpha\chi_1(t) = a(t)\chi_L\{\xi_i\}$ and $\Psi_{1,2}$ independent of time. Now, the crucial point is that even though the overlap of the environment wave functions $\chi_L\{\xi_i\}$ and $\chi_R\{\xi_i\}$ may be very small compared to unity, this does not destroy the quantum coherence behavior — all it does is to lengthen the characteristic time-scale (period of oscillation) considerably by comparison with that for the isolated system. In fact, as we shall see in Sec. 5, the frequency of oscillation between the two wells is reduced (by the overlap factor $|\langle\chi_L\{\xi_i\}|\chi_R\{\xi_i\}\rangle|$), but if we wait long enough we shall get back to our original conditions ($a(t) = 1$, $b(t) = 0$), and the value of $P_L(t)$ will at that time be $+1$ just as in the isolated case. The feature of the problem of which we are implicitly taking advantage, here, is that in this experiment we do not measure any "interference terms" of the form $\langle\Psi_L|\Omega|\Psi_R\rangle$ where Ω is an operator referring only to the system, and hence the fact that $\chi_L\{\xi_i\}$ and $\chi_R\{\xi_i\}$ are nearly orthogonal does not hurt us. (As discussed in Leggett (1980 and 1984a), in effect this experiment measures the matrix element $\langle\Psi_L|\exp(-i\hat{H}t)|\Psi_R\rangle$, but the Hamiltonian $\hat{H}$ acts on both the system *and* the environment, so the objection does not apply.)

Before concluding that the adiabatic part of the system-environment coupling is harmless to quantum coherence, one might worry about the following question. We have said that the reduction of the oscillation frequency is proportional to the (small) quantity $\langle\chi_L\{\xi_i\}|\chi_R\{\xi_i\}\rangle$. Now, in practice the experiment is conducted repeatedly on the same system, and the physically meaningful quantity $P_L(t)$ is actually an average of the results 1 or 0 obtained on each specific run. But we cannot control the initial state of the environment perfectly, so we might worry that the above factor is different on each run. In that case we have in effect not a single ensemble, but an average over many different ensembles with different (predicted) oscillation frequencies, so the total $P_L(t)$ should average to zero for all the relevant t and we expect nothing interesting. While this argument is not incorrect, it fortunately turns out that in practice the uncertainty in $\langle\chi_L|\chi_r\rangle$ (given adequate control of the external parameters, of course) is of order $N^{-1/2}$ relative to this quantity itself, where N is the number of particles in the environment, so the effect just considered is of negligible magnitude. More generally, we can argue (cf. Caldeira and Leggett (1983) p. 444) that in a macroscopic system "intrinsic" random fluctuations in the behavior of the environment do not by themselves spoil the coherence behavior. This point, incidentally, seems not to have been universally appreciated in the literature on the quantum theory of measurement, where the impression is often given that such fluctuations are automatically sufficient to destroy coherence at the macroscopic level.

If, therefore, adiabatic effects are "harmless" in the sense implied, it remains to discuss the effect of the non-adiabatic terms, i.e. these coupling the system to those modes of the environment which have frequency comparable to, or less than, these of the system ($\sim \omega_0$). As regards modes with frequencies far below ω_0, these are to all intents and purposes inert: they simply contribute to the universe wavefunction an overall multiplying factor which clearly has no role in destroying coherence. Thus the "dangerous" modes are those with frequencies of the order of ω_0. But these are precisely the modes which can exchange energy irreversibly with the system and hence act, *inter alia,* to damp its quasi-classical motion (we will confirm this in detail in the context of MQT and MQC in the next sections). Thus we reach a conclusion whose importance for our general program of research cannot be overestimated: *the part of the system-environment interaction which tends to destroy quantum coherence is just the part which causes dissipation in the classical motion.* Dissipation, and the destruction of coherence, are the classical and quantum sides of the same coin! (This point is discussed further in Leggett (1984a)).

This suggests the following question: is it possible to formulate a general, *quantitative* relationship between the classical dissipation observed in a given system and the degree of destruction of coherence (in, say, an MQC experiment) in that same system? At present no way to do so rigorously for a quite general system is known. However, once we are prepared to make a certain postulate about the form of the system-environment interaction, which though not totally rigorous is at least extremely plausible, then we can construct such a quantitative relationship, and this has been the basis of most of the work in this area in the last few years.

The postulate in question is that, *for the purposes of describing its effect on the behavior* (either classical or quantum mechanical) *of the system,* the environment may be described as a set of noninteracting linear harmonic oscillators, and moreover that the system-environment coupling may be taken to be linear in the oscillator coordinates. Thus, if we write the system coordinate generically as q and that of the i-th oscillator as x_i, and call the corresponding momenta p and p_i respectively, the Hamiltonian of the system plus environment is postulated to be of the form

$$H = \frac{p^2}{2M} + V(q) + \sum_i \left(\frac{p^2}{2m_i} + \frac{1}{2} m_i \omega_i^2 x_i^2 \right)$$

$$- \sum_i f_i(q) x_i + \sum_i \frac{f_i^2(q)}{2m_i \omega_i^2}, \qquad (3.7)$$

where the last term is conventionally introduced for technical convenience (see Caldeira and Leggett (1983) Sec. 2 and appendix A).[18]

That the Hamiltonian in Eq. (3.7) is an adequate description of the interaction of any macroscopic system with its environment (for the purpose stated) is not at all obvious at first sight. A detailed argument for this claim, at zero temperature, is given in appendix C of Caldeira and Leggett (1983)), for the case where the classical dissipation is known to be proportional to $\dot{q}^2$, and this is generalized, in Leggett (1984c), to the case where the dissipation (and the reactive effects of the environment) have an arbitrary frequency dependence but the system-environment interaction is known *a priori* to be linear in q ("strictly linear" dissipation); in that case we obviously have $f_i(q) \equiv qC_i$ in Eq. (3.7). Even for these cases the argument is not totally rigorous, and it has not been explicitly generalized either to the case where the dissipation has an arbitrary dependence on both amplitude and frequency or to finite temperatures (a lacuna which has not prevented the Hamiltonian in Eq. (3.7) being used as a general description of dissipation under these extended conditions in a growing number of papers in the literature). Despite the absence of a totally rigorous foundation for it, it is nevertheless tempting to believe that Eq. (3.7) is in fact an adequate description of the interaction of any macroscopic system with its environment and that the loopholes mentioned above could in principle be filled (the urgency of filling them will of course depend crucially on the progress of experiments to observe MQC, cf. above). This belief is reinforced by the fact that (a) most plausible microscopic models of specific macroscopic systems interacting with their environments (e.g. in laser physics or solid-state physics) lead to Hamiltonians which are explicitly of the form in Eq. (3.7), and (b) in the few cases (Ambegaokar *et al.*, 1982; Eckern *et al.*, 1984; Chang and Chakravarty, 1985) where the models studied are not explicitly of this form, the functional integrals which have been obtained from them, at least in the parameter regimes important for MQT and MQC, are virtually identical to those which would be obtained from Eq. (3.7) with a suitable choice of parameters; this strongly suggests (though it does not rigorously prove) that the Hamiltonian in these cases could in fact be cast in the form in Eq. (3.7) by an appropriate choice of environment variables. Certainly, more research into this question is desirable.

Once given the Hamiltonian in Eq. (3.7), it is relatively straightforward to integrate out the environment variables and obtain expressions for the behavior of the system. In the case of quasiclassical motion this can be done by writing down equations of motion for q in terms of x_i and vice versa, and eliminating

---

[18]M is the generalized "inertia" of the system and need not be a real mass. For example, in the case of a SQUID ring it is the junction capacitance C (cf. Eq. (2.29)). The m_i are similarly the fictitious "masses" of the oscillators.

the x_i (see Caldeira and Leggett (1983), appendix C). In general the result is quite complicated in form (see Eq. (C. 39) of Calderia and Leggett (1983)). However, it simplifies enormously if the system-environment interaction is not only linear in the x_i but also linear in q. Although there is no general *priori* reason why this should be so, it fortunately turns out to be a very good approximation in almost all cases of physical interest, either because of the known general nature of the interaction (e.g., in the case of a Josephson junction shunted by a physical resistor, the interaction is the standard $\mathbf{p} \cdot \mathbf{A}$ term of quantum electrodynamics and hence is known to be linear in the flux Φ, see Appendix A of Caldeira and Leggett (1983)) or because one is interested in a sufficiently small region of q that a Taylor expansion in q up to the first-order term is adequate (as in the case of MQT in a tunnel-oxide junction, *ibid.* p. 397). In all these cases one can legitimately make in Eq. (3.7) the substitution

$$f_i(q) = q C_i \quad . \tag{3.8}$$

The resultant Hamiltonian has been widely used in the literature for the general discussion of the effect of the environment on MQT and MQC; one should, however, bear in mind that some of the results obtained from it are not valid for the more general case (which can be handled by the same techniques (see Appendix C of Caldeira and Leggett (1983)). In particular, the unique relation obtained in Sec. 4 below between the classical dissipation and the MQT rate becomes an inequality (*ibid.*, Eq. (4.40)).

For the special case (3.8) of the Hamiltonian (3.7) it turns out that there is a single function which completely encapsulates the effect of the environment on the system[19], namely the quantity

$$J(\omega) \equiv \frac{\pi}{2} \sum_i \frac{C_i^2}{m_i \omega_i} \delta(\omega - \omega_i) \quad . \tag{3.9}$$

Environments having identical forms of $J(\omega)$ are identical in their effects on the system. It is actually more convenient to work in terms of a related quantity $K(\omega)$ defined by[20]

$$K(\omega) \equiv -\omega^2 \left[M + \frac{2}{\pi} \int_0^\infty \frac{J(\omega')d\omega'}{\omega'(\omega'^2 - \omega^2)} \right] . \tag{3.10}$$

[19] This statement is strictly true only if the initial conditions on the environment are specified in some standard way (e.g., as corresponding to thermal equilibrium), as is always the relevant condition in practice.

[20] This definition follows that of Leggett (1984c). (See Eq. (2.13) of that reference: $\overline{K}(\omega)$ is shown there to be identical to the $K(\omega)$ defined in Eq. (1.1).) The quantity $K(\omega)$ defined in Caldeira and Leggett (1983) is different by a few trivial factors.

It then follows straightforwardly (Leggett, 1984c) that in the quasiclassical motion the system trajectory $q(t)$ has a Fourier transform $q(\omega)$ which satisfies the equation of motion

$$K(\omega)q(\omega) = -\left(\frac{\partial V}{\partial q}\right)(\omega) \quad , \tag{3.11}$$

(where $(\partial V/\partial q)(\omega)$ means the Fourier transform of the quantity $(\partial V/\partial q)(t)$ with $\partial V/\partial q$ having the value appropriate to the trajectory $q(t)$). Thus, a complete knowledge of the classical equation of motion will automatically determine $K(\omega)$. As we shall see in the next two sections, a knowledge of $K(\omega)$ in turn uniquely determines the MQT and MQC behavior. Thus, within the model described by the Hamiltonian (3.7) with the substitution (3.8), *the classical equation of motion uniquely determines the behavior in an MQT or MQC experiment.* Of course, this conclusion needs to be tempered by the observations that (a) the argument that the model in question applies to all systems of physical interest is not completely rigorous (see above), and (b) as we shall see in the next section, the business of inferring the complete classical equation of motion from practical experiments may not be all trivial. Nevertheless, the above conclusion takes us a long way towards our goal of a *model-independent* test of the validity of quantum mechanics at the macroscopic level.

4. Macroscopic Quantum Tunneling

The problem of macroscopic quantum tunneling (MQT) is defined as follows. We consider a system described by some variable q such that "appreciably different" values of q (cf. below) correspond to states which are by some reasonable criterion macroscopically different. We imagine that the potential energy $V(q)$ which is associated with the variable q has a metastable minimum which is separated from a region of considerably lower potential energy by a barrier of height U_0 and width Δq (see Fig. 5). The region outside the barrier is regarded as unbounded, i.e., $V(q) < 0$ for all $q > \Delta q$. We then imagine that the variable q is known at some initial time to have a value within the metastable well, and we ask for the probability, at some later time t, that the system is still to be found within the well. This is the exact analog of the problem of tunneling of an α-particle out of a heavy nucleus, and, guided by the analogy, we should expect that for not too short and not too long times the probability would decay approximately exponentially with a decay constant which we call Γ. However, in the case of quantum tunneling of a macroscopic variable there are

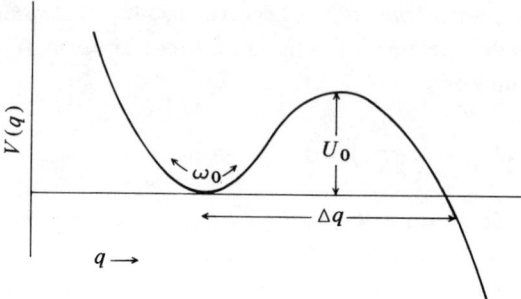

Fig. 5. Macroscopic quantum tunneling.

a number of features which differ from the α-decay case: (1) If "appreciably different" refers to values of q differing by an amount of order Δq, then auto-matically the tunneling process takes place between macroscopically different states, (and so observation of it constitutes some circumstantial evidence that quantum mechanics is still working at the macroscopic level; this, of course, is why we are interested in it). (2) As we have emphasized above, a macroscopic variable is inevitably coupled to its environment, and it is essential to take this coupling into account when calculating the decay rates. (Strictly speaking, an α-particle attempting to escape from a nucleus is also coupled to an "environ-ment" — the internal modes of vibration of the nucleus — and recent calculations in nuclear physics do take account of this, but because of the relatively small number of degrees of freedom the situation is rather different from the one considered here.) (3) Because the characteristic frequency of the macroscopic variable is usually low ($\omega \sim M^{-1/2}$, where M is the associated inertia) it is relatively easy to get the temperature high enough that the system escapes from the well by a classical thermal fluctuation process rather than by quantum tunneling. Thus it is of interest to calculate the temperature dependence of the decay rate. (By contrast, in α-particle decay the characteristic frequencies are so high ($\sim \mathrm{MeV}/\hbar$) that even the highest laboratory temperatures are effectively zero in the content of α-decay. However, there are of course many similar problems in chemical and solid-state physics when the temperature-dependence of the escape rate is nontrivial.)

Let us consider explicitly the case of an rf SQUID ring. In this case the macroscopic variable in question is the trapped flux Φ, and, if we could neglect the coupling to the environment the potential energy associated with it would be of the form (2.16), that is

$$U(\Phi) = \frac{(\Phi - \Phi_x)^2}{2L} - \frac{1}{2\pi} I_c \Phi_0 \cos\left(2\pi \frac{\Phi}{\Phi_0}\right) . \tag{4.1}$$

Here Φ_x is the externally applied flux, which we treat as a c-number, I_c and L are respectively the critical current of the Josephson junction and the self-inductance of the ring, and Φ_0 is the flux quantum $h/2e$. Moreover, as we explained in Sec. 2, the "kinetic energy" has the form

$$T = \frac{1}{2} C\dot{\Phi}^2 = \frac{1}{2C} p_\Phi^2 \quad , \tag{4.2}$$

so that the capacitance C of the junction plays the role of a particle "mass". As we noted, for a value of $\beta_L \equiv 2\pi L I_c/\Phi_0$ greater than one the potential in Eq. (4.1) has one or more metastable minima. Strictly speaking, the situation here does not correspond exactly to the one discussed above, since the region "outside the barrier" is never infinite; however, for $\beta_L \gg 1$ it is so large in relation to the "inside" region that it should be good approximation to regard it as infinite. (A more careful discussion of this question is given in Caldeira and Leggett (1983), pp. 420-1.) For β_L close to 1 the situation corresponds rather to the "macroscopic quantum coherence" setup described in the next section.

Most experiments on MQT have actually been done on current-biased junctions rather than on SQUID rings. The variable is now the phase difference $\Delta\varphi$ between the condensate wave function on the two sides of the junction, which is strictly defined only modulo 2π. If we ignore the complications associated with this feature (see the discussion at the end of Sec. 2) then the situation is identical to that of a SQUID ring in the limit $\beta_L \to \infty$ with the replacement $\Phi \to (\Phi_0/2\pi)$ $\Delta\phi$, $\Phi_x \to LI_x$, where I_x is the externally imposed current (cf. Eq. (2.41)). Thus the results for a SQUID ring (in the limit $\beta_L \to \infty$) can be trivially transposed to the case of a biased junction and vice versa. Below I quote the relevant formulas explicitly for a biased junction. We are usually interested in the case when I_x is close to the critical current I_c, so that the height of the barrier is small compared to $I_c\Phi_0/2\pi$. In that limit it is straightforward to show that the barrier height U_0 and the small-oscillation frequency $\omega_0 \equiv [(C\Phi_0^2/4\pi^2)^{-1}$ $(\partial^2 U/\partial\phi^2)_{\min}]^{1/2}$ (when the subscript "min" means that the derivative is evaluated at the position of the metastable minimum) are given by the formulas (cf. Leggett (1984b)

$$U_0 = \frac{2\sqrt{2}}{3\pi} \Phi_0 I_c (\delta I/I_c)^{3/2} \quad , \tag{4.3}$$

$$\omega_0 = (2\pi I_c/C\Phi_0)^{1/2} (2\delta I/I_c)^{1/4} \quad , \tag{4.4}$$

where $\delta I \equiv I_c - I_x$. Moreover, in this limit the shape of the barrier is well approximated by a cubic, i.e., (putting $\Delta\phi \equiv q$ and writing the width of the barrier as Δq)

$$U(q) = \frac{3}{2} U_0 \left\{ \left(\frac{q}{\Delta q} \right)^2 - \left(\frac{q}{\Delta q} \right)^3 \right\} . \tag{4.5}$$

For this form of potential, if the system could be regarded as isolated, the quantum tunneling rate at zero temperature can be straightforwardly calculated by the WKB technique, with the result in Eq. (2.39).

Our problem, however, is to obtain the rate of decay (assuming it is well-defined) in the presence of coupling to the environment. At this point we have a choice of alternative strategies. We can start as it were from scratch, that is from a specific microscopic model of the system and its environment, and calculate the decay rate directly, without the need to go through the arguments of Secs. 2 and 3. This is the strategy of Eckern *et al.* (Ambegaokar *et al.*, 1982; Eckern *et al.*, 1984), with results which will be described below. The alternative is to use the technique described in Sec. 3 to model the system-environment interaction, extracting the necessary combination of parameters (i.e., the function $K(\omega)$ or its generalization) directly from the experimental data on quasiclassical motion and to calculate the tunneling rate from the Hamiltonian in Eq. (3.7). It is possible to do this for an arbitrary form of the interaction function $f_i(q)$ (see Caldeira and Leggett (1983), Sec. 4), but in this case it is usually not easy to extract the form of $f_i(q)$ from the quasiclassical behavior. Fortunately, as we saw above, in many cases of practical interest it is possible to replace $f_i(q)$ by the simple form qC_i, and in this case $K(\omega)$ can, at least in principle, be uniquely extracted from the classical data. I therefore confine the explicit discussion to this case, which is the one overwhelmingly studied in the literature.

We start, then, from the Hamiltonian in Eq. (3.7) with the replacement $f_i(q) \rightarrow qC_i$, that is

$$H = \frac{p^2}{2M + V(q)} + \sum_i \left(\frac{p_i^2}{2m_i} + \frac{1}{2} m_i \omega_i^2 x_i^2 \right)$$

$$- q \sum_i C_i x_i + q^2 \sum_i \frac{C_i^2}{2m_i \omega_i^2} , \tag{4.6}$$

and wish to calculate the tunneling rate (for the moment at zero temperature) out of the metastable minimum, which evidently occurs at $q = 0$, $x_i = 0$. In principle it would be possible to do this by the appropriate generalization (Kapur and Peierls, 1937; Landauer, 1950; Banks *et al.*, 1973) of the WKB technique to many degrees of freedom. However, there exists one technique which is particularly convenient for our purpose, namely the "bounce" ("instanton") technique which was originally formulated by Langer[21] (1967) in the content of first-order phase transition theory and rediscovered in a particle-physics context by Stone (1977) and by Callan and Coleman (1977). In the context of a different but related problem, namely the tunneling between two *degenerate* potential minima, the relationship between functional-integral ("instanton") methods and the many-dimensional WKB approach has been studied by De Vega *et al.* (1979), who conclude that the two methods give identical results for this problem. In a very recent paper, Schmid (1986) has given a careful discussion of the WKB method for the many-dimensional decay problem (a special case of which is described by the Hamiltonian in Eq. (4.6)). In this paper Schmid demonstrates that the adaptation of the simplest version of the one-dimensional WKB method to the many-dimensional case gives results in exact agreement with those obtained by the instanton method; however, he also notes that there are nontrivial differences between the one- and many-dimensional problems, and in particular that for certain forms of the Hamiltonian, which include that of Eq. (4.6) with the constraint $J(\omega) = \eta \omega$ (cf. Eq. (3.9), and below), it is not obvious that a straightforward adaptation of the conceptually most satisfactory version of the one-dimensional technique is possible. While it seems very unlikely that this feature would affect the exponent of the WKB expression for the tunneling rate, it is not entirely obvious that the prefactor calculated by the simple version of the method (which, as noted above, is identical to that obtained by the instanton method) is totally reliable. This question, which is at the time of writing to the best of the author's knowledge unresolved, should be borne in mind when considering both the detailed fit of theory and experiment and the generalization of the theory to finite temperature (cf. below).

In the instanton technique, one studies the density matrix of the "universe" (system plus environment) at inverse temperature $\beta \equiv (k_B T)^{-1}$:

$$\rho(q, q' : \{x_i\}, \{x_i'\} : \beta) \equiv \sum_n e^{-\beta E_n} \psi_n^*(q, \{x_i\}) \psi_n(q', \{x_i'\}) ,$$

$$(4.7)$$

where the ψ_n are the exact eigenstates of the universe. As $T \to 0$ the only contribution comes from the groundstate, $n = 0$, so that irrespective of the

---

[21] Langer's paper is itself one step in a long history of work on the classical activation problem in many dimensions, see, e.g., Buttiker and Landauer (1981, 1983). In the present context its importance is that it was (to the best of my knowledge) the first to formulate the quantum decay problem in functional-integral terms.

other arguments ρ is proportional to $\exp(-\beta E_0)$. One then represents ρ by a path integral (see Feynman and Hibbs (1965)).

$$\rho(q, q':\{x_i\},\{x_i'\}:\beta) = \int \mathscr{D}q(t) \int \prod_i \mathscr{D}x_i(t) \exp -S[q, \{x_i\}]/\hbar \ ,$$
(4.8)

with the constraint on the paths that $q(0) = q, x_i(0) = x_i, q(\beta) = q', x_i(\beta) = x_i'$, and where $S[q(t), x_i \ (t)]$ is the "Euclidean action", that is, the quantity

$$S[q(t),\{x_i\} \ (t)] = \int_0^\beta dt \, \mathscr{L}[q, \dot{q}, \{x_i\}, \{\dot{x}_i\}] \ ,$$
(4.9)

where in the case of Hamiltonians such as in Eq. (4.6) with no velocity-dependent forces the integrand is just the Hamiltonian expressed in the variables $q, \dot{q}, x_i, \dot{x}_i$ rather than q, p, x_i, p_i. Now this quantity is just the Lagrangian for a (many-dimensional) system with coordinates $q(t), x_i(t)$ moving in the *inverted* potential $\tilde{V}(q,\{x_i\}) \equiv -V(q,\{x_i\})$. It therefore follows immediately that its extrema occur on those trajectories $(q(t), x_i(t))$ which correspond to classical motion of the system in this potential. It is clear that the functional integral in Eq. (4.8) will be dominated by paths near that one (or ones) which make the action in Eq. (4.9) an absolute minimum. If the action on this path is S_0, and there are no others, then the integral in Eq. (4.8) will be proportional to $\exp(-S_0/\hbar)$, the prefactor being determined from the small fluctuations around this path.

The point of the instanton technique is that in an unstable system there exists at least one point,[22] other than the origin, at which $V(q, x_i) = 0$, and therefore there must be at least one classical trajectory in the inverted potential in which the system rolls out from the origin to this point and back. This path is called a "bounce", or more loosely an "instanton." Because the (actual) potential beyond this point is negative, it turns out that this is not a minimum but a *saddlepoint* of the action, and therefore the small fluctuations around it give a divergent prefactor. At this point one carries out a mathematical trick, which it is difficult to justify rigorously. One deforms the path into the complex plane, obtaining thereby a contribution to the action which is pure imaginary, and then proceeds to identify this with an imaginary contribution to the groundstate energy E_0. Since, in real time, an imaginary part of the energy is equivalent to an exponential decay in time, one finally expresses the decay rate Γ of the metastable rate in terms of the contribution to the action of the "bounce" path. The details of this procedure are explained,

---

[22] In fact in a many-dimensional system there is a whole surface having this property.

e.g., in Callan and Coleman (1977) and Coleman (1979) (for the simple one-dimensional case); one's confidence in it is increased by the fact that it gives exactly the same result, for both the one-dimensional and the many-dimensional case, as the standard WKB procedure (through cf. above). The final result is

$$\Gamma = A \exp(-B/\hbar) \quad , \tag{4.10}$$

where B is the classical action (in the inverted potential) evaluated along the "bounce" path in the many-dimensional space, and A is a prefactor which is given by a ratio of determinants corresponding to the fluctuations around this path and around the "trivial" path $q(t) = x_i(t) = 0$.

It is possible, in principle, to use this technique for our problem as it stands (cf. Schmid (1986)). However, a very great technical simplification is achieved if we decide to calculate explicitly not the complete density matrix of the universe, Eq. (4.7), but the reduced density matrix $\rho(q, q' : \beta)$ of the system, that is, the quantity obtained by putting in Eq. (4.7) (or Eq. (4.8)) $x_i' = x_i$ for all i and integrating over all the x_i. Since we are interested, in the end, only in the form of the groundstate energy and not in the wave functions, this should essentially give the same information. Moreover, it has the advantage that since the Hamiltonian contains no terms higher than bilinear in x_i, both the functional integrals over $x_i(t)$ in Eq. (4.8) and the ordinary integrals over x_i can be done analytically. When this is done, the quantity $\rho(q, q' : \beta)$ is expressed as a functional integral over one-dimensional paths $q(t)$, with an action which now contains an extra term of the form $\alpha(t - t')(q(t) - q(t'))^2$ which expresses the effect of the eliminated environmental oscillators. In fact, with a little algebra (see Caldeira and Leggett (1983) and Leggett (1984c)) it can be shown that the expression for $\rho(q, q' : \beta)$ is

$$\rho(q, q' : \beta) = \int \mathscr{D}q(t) \exp(-S_{\text{eff}}[q(t)]/\hbar) \quad , \tag{4.11}$$

where the boundary conditions are $q(0) = q$, $q(\beta) = q'$, and the effective action $S_{\text{eff}}[q(t)]$ is most conveniently expressed in terms of the Fourier transform $q(\omega)$ of $q(t)$, i.e.,

$$S_{\text{eff}} = \left\{ \frac{1}{2\pi} \int_{-\infty}^{\infty} K(-i|\omega|)|q(\omega)|^2 d\omega + \int_{-\infty}^{\infty} V[q(t)] dt \right\} . \tag{4.12}$$

(The form (4.12) is appropriate to the case $\beta = \infty$ ($T = 0$). At finite T the first integral would be replaced by a sum over $\omega_n = 2\pi n/\beta$ and the second would run from 0 to β, cf. below.) Here $K(\omega)$ is just the function which defined the *classical* motion of the system (see Eq. (3.11)). The result in Eq. (4.11), with Eq. (4.12), is formally exact. Once it is obtained, we proceed exactly as above and obtain a tunneling rate of the form (4.10), where as before B is the saddlepoint value of the effective action in Eq. (4.11) and A is a ratio of determinants corresponding to the fluctuations around the corresponding "bounce" path and around the trivial path $q(t) = 0$. (See Caldeira and Leggett (1983) for details). The great advantage is that we now have only a one-dimensional problem and there is considerable hope of obtaining an accurate analytical or numerical value of the exponent B and prefactor A. It has been demonstrated explicitly by Schmid[23] (1986) that the results obtained from Eqs. (4.11) and (4.12) are identical to those obtained by the many-dimensional WKB method (though see above).

The upshot of all this is that, within the context of our model, we can now calculate tunneling rates uniquely in terms of the function $K(\omega)$ which in principle can in turn be found from the experimentally observed classical motion. A case of particular interest arises if we can assume that the quantity $K(\omega)$ can be expanded in a Taylor series around its $\omega = 0$ value. This will be so, crudely speaking, if the characteristic frequency of the environment is much larger than the frequencies of interest in the tunneling problem, and if the system itself has no "hidden" degrees of freedom on the scale of those frequencies (cf. below). In that case we write

$$K(\omega) = K(0) + \omega K'(0) + \tfrac{1}{2}\omega^2 K''(0) + 0(\omega^3) \ . \qquad (4.13)$$

The zeroth-order term can be lumped in with the potential $V(q)$ as an extra contribution to the quadratic term. The first-order term obviously corresponds[24] to a friction coefficient $\eta = -iK'(0)$ (or in the Josephson junction problem, to a shunt conductance $\sigma_n = -iK'(0)$), and the second-order term to a mass $M = -K''(0)$ (or, in the junction case, a capacitance $C = -K''(0)$). Thus the effective action in Eq. (4.11) reads

$$S_{\text{eff}} = \frac{1}{2\pi} \int_{-\infty}^{\infty} (M\omega^2 + \eta|\omega|)|q(\omega)|^2 \, d\omega + \int_{-\infty}^{\infty} V[q(t)] \, dt \ .$$
$$(4.14)$$

The problem of finding the tunneling exponent B and prefactor A for the specific form of action in Eq. (4.14) at $T = 0$ has been treated in detail by analytical methods in Caldeira and Leggett (1983) and numerically in Chang

---

[23] Schmid's calculation (Sec. 9 of his paper) considers a particular form of $J(\omega)$ (Eq. (9.2)), hence of $K(-i|\omega|)$: but it is clear that the proof goes through for the general case.

[24] Note that time-reversal invariance implies that terms odd in ω must be pure imaginary.

and Chakravarty (1984). If the potential $V(q)$ is written in the convenient form

$$V(q) = \tfrac{1}{2}M\omega_0^2(q^2 - q^3/\Delta q) \ , \tag{4.15}$$

and we define $\alpha \equiv \eta/2M\omega_0$, then the result for the rate Γ can be written succinctly in the form

$$\Gamma = f(\alpha)\omega_0\left[(216U_0/\pi\hbar\omega_0)^{1/2}\right]\exp-\left[\frac{36}{5}\left\{1 + \frac{15}{4}\,\alpha A_0(\alpha)\right\}\frac{U_0}{\hbar\omega_0}\right] \tag{4.16}$$

where $A_0(\alpha)$ is a slowly varying function of α which tends to $12\zeta(3)/\pi^3 \cong$ 0.47 in the limit $\alpha \to 0$ and to $2\pi/9 \cong 0.70$ in the limit $\alpha \to \infty$. (It is amusing to note that for many purposes the quantity $(\omega^2 - (\eta/2M)^2)^{1/2} - \eta/2M \equiv \omega_{eff}$ is a characteristic frequency of the damped classical motion: if we write the WKB exponent as $z(\alpha)U_0/\hbar\omega_{eff}$, then $z(\alpha)$ is constant to within about 30% over the whole range of α.) The quantity $f(\alpha)$ tends to 1 as $\alpha \to 0$, and to a constant times $\alpha^{7/2}$ as $\alpha \to \infty$.

It is possible to analyze cases when $K(\omega)$ has a more complicated structure in a similar way. Thus, in principle, in should be possible to predict with high numerical accuracy the tunneling rate in any Josephson junction whose classical behavior is known.

An alternative approach to the problem of tunneling in Josephson junctions and SQUID rings was taken by Eckern *et al.* (Ambegaokar *et al.*, 1982; Eckern *et al.*, 1984). Instead of trying to relate the rate to the classical motion, they started from a well-defined microscopic model (the BCS model of superconductivity in bulk plus the Bardeen-Josephson tunneling Hamiltonian) and calculated the effective action $S_{eff}(\Delta\phi)$ directly. In the limit $T \to 0$, and provided one can assume that all frequencies of relevance are small compared to the gap frequencies 2Δ, their result takes a strikingly simple form. The action is just the one which one would get for an *isolated* junction, i.e., from the Schrödinger equation (Eq. (2.29)), but with a correction to the capacitance of the form

$$\delta C = C_0(1 - \tfrac{1}{3}\cos\Delta\phi) \ , \tag{4.17}$$

$$C_0 \equiv 3\pi\hbar/(32\Delta R_N C) \ , \tag{4.18}$$

when Δ is the bulk energy gap and R_N the resistance of the junction in the normal state. (For all except the smallest junctions, the correction in Eq. (4.17) is at most a few percent.) In the light of hindsight this result is not very surprising. In the tunneling-Hamiltonian model the only possible mechanism of dissipation is associated with the normal component, which is not there at $T = 0$, so there is nothing corresponding to a resistance in the problem, and the only effect of coupling to the "environment" (in this case the virtual normal component) is reactive; since the excited states to which the motion of the phase could couple (particle-hole pair states) are pushed up in the super-conducting state, the frequency of the motion itself is pushed down in the usual way, corresponding to an increase in effective capacitance.

We now turn to the question of the escape rate at finite temperature. The situation here is conceptually rather more complicated than at zero temperature. Suppose that we could somehow prepare the "universe" (or rather, an ensemble of "universes") so that the top of the barrier was replaced by a rigid wall, and allow it to come into equilibrium under these conditions; then the various energy levels will be populated according to the Boltzmann factor $\exp(-\beta E_n)$. If now we imagine removing the barrier at $t = 0$, then the *initial* decay rate is plausibly

$$\Gamma_0 \equiv \frac{dP(t)}{dt}\bigg|_{t=0+} = \sum_n \Gamma_n e^{-\beta E_n} \bigg/ \sum_n e^{-\beta E_n} \quad , \tag{4.19}$$

where Γ_n is the decay rate of the n-th level. Now, the free energy of the system before removal of the partition is given by the expression

$$F = -\beta^{-1} \ln \sum_n e^{-\beta E_n} \quad . \tag{4.20}$$

Once the partition is removed, the energy levels are not quite well-defined, because of the possibility of escape to the continuum. It is however tempting to try to take this into account by assigning to each level a *complex* energy, just as we did to the groundstate in the $T = 0$ calculation; thus we make the replacement

$$E_n \rightarrow \tilde{E}_n \equiv E_n + \tfrac{1}{2} i \hbar \Gamma_n \quad . \tag{4.21}$$

Now, if we redefine F so that E_n in Eq. (4.20) is replaced by $\tilde{E}_n$, we find that in the limit $\Gamma_n \ll E_n$ we have, formally, the relation

$$\Gamma_0 = \frac{2}{\hbar} \operatorname{Im} F \quad . \tag{4.22}$$

If, moreover, the population of the n-th level is maintained so as to be propor-tional to $\exp(-\beta E_n)$ for all t, then we expect

$$P(t) = \exp(-\Gamma t), \qquad \Gamma \equiv \Gamma_0 \quad , \tag{4.23}$$

so that Eq. (4.19) refers not just to the initial decay rate to the steady-state rate. Furthermore, if we represent $F \equiv -\beta^{-1} \ln \mathrm{Tr}\, \hat{\rho}$ as a functional integral and integrate out the environment variables as at $T = 0$, we have

$$F = -\beta^{-1} \ln \int dq\, \rho(q, q : \beta) \quad , \tag{4.24}$$

where the system reduced density matrix $\rho(q, q':\beta)$ is given by an expression identical to Eq. (4.11) except for the replacements noted below Eq. (4.12). As in the $T = 0$ case, we can identify at least one trajectory which is a *saddle-point* rather than a minimum of the effective action, and it is tempting to identify the imaginary part of F with the contribution from this path. If we do this, then we finally obtain a finite-temperature tunneling rate of a form identical to Eq. (4.10), but with B now the saddlepoint value of the action functional in Eq. (4.12) (written now in the form appropriate to finite temperature)

$$S_{\mathrm{eff}} = \sum_n K(-i|\omega_n|)|q(\omega_n)|^2 + \int_0^{\beta\hbar} V[q(t)]\, dt \quad , \tag{4.25}$$

subject to the condition $q(0) = q(\beta)$ ($\neq 0$ in general). The path which gives the saddlepoint value is just the periodic classical path in the "inverted potential" which has period $\tau \equiv \beta\hbar$. The prefactor A is similarly determined by the small fluctuations around this path. Thus the problem is reduced, as at $T = 0$, to the purely mathematical problem of finding this path and calculating the action along it and near it.

The above prescription[25] has been used in a number of calculations (Affleck, 1981; Larkin and Ovchinnikov, 1983a, 1983b, 1984; Grabert *et al.,* 1984, 1985; Grabert and Weiss, 1984, 1984b; Hanggi *et al.,* 1985; Riseborough *et al.,* 1985) of the finite temperature escape rate in the tunneling regime, in particular by the Moscow and Stuttgart-New York Polytechnic groups (cf. also Wolynes, 1981; Mel'nikov and Meshkov, 1983; Weiss *et al.,* 1984)). It is clear that the "justification" given above raises questions additional to the ones already present at zero temperature; in particular the notion of working with exact energy eigenvalues and yet keeping them continuously repopulated so that the population is always proportional to the Boltzmann factor seems a rather delicate one. Other justifications of Eq. (4.22) given in the literature also seem (to the present author at least) less than totally convincing. An alternative approach to the problem was given by Waxman (Waxman, 1984; Waxman and

---

[25] With appropriate modifications at higher temperatures, cf. below.

Leggett, 1985); this explicitly assumes that the decay rate is proportional to the equilibrium probability density which one would have of finding the system just inside the far edge of the barrier if a rigid wall were placed at the edge, and then calculates the latter by a functional integral technique. Although this method also to some extent begs the question about "continuous repopulation", it does at least avoid having to deal in complex free energies. The result of this method is that the expression for B is exactly equivalent to that obtained in the above ("instanton") method, but that the prefactor is not obviously the same. Further elucidation of this question, perhaps by a generalization of Schmid's approach (Schmid (1986)) to finite temperatures, seems called for; however, there seems little room to doubt that the WKB exponent is correctly given by either of these methods, even if the prefactor is not. Since the WKB limit the variation of the prefactor is usually a fairly minor effect, except in very heavily damped systems, this is more or less adequate to our purpose.

In any case, it is clear that neither method as it stands can be extrapolated to arbitrarily high temperatures. This is because, for any specific shape of potential and form of $K(\omega)$, there is a minimum period, say $2\pi/\omega_m$, for a periodic trajectory in the inverted potential. Since the instanton path is subjected to the constraint of being periodic with period $\beta \hbar$, it is clear that at temperatures higher than the temperature T_0 defined by

$$kT_0 = \hbar\omega_m/2\pi \quad , \tag{4.26}$$

there is no instanton path. In fact, we should expect that at sufficiently high temperatures the escape from the metastable wall would be dominated, not by quantum tunneling through the barrier, but by thermal activation above the barrier, which would give a rate proportional to the classical Gibbs probability $\exp(-U_0/kT)$ to find the system with energy above the barrier top. This process was calculated in a classic paper by Kramers (1940) whose results are generally believed today. It is, of course, highly desirable to find a general technique which would not only give the correct tunneling rate for $T < T_0$, but also correctly give the classical result for $T \gg T_0$ and the combined effects of classical activation and tunneling in the region $T \sim T_0$. A number of papers in the literature (Larkin and Ovchinnikov, 1984; Grabert *et al.*, 1984, 1985; Grabert and Weiss, 1984a, 1984b; Hanggi *et al.*, 1985; Riseborough *et al.*, 1985) have attempted to formulate such a technique. However, to the best of the author's knowledge all attempts so far[26] in this direction have suffered from one major defect: in the classical ("Kramers") limit $T \gg T_0$ they all

---

[26] See however note at end of section.

predict a rate of the form

$$\Gamma \sim \frac{\widetilde{\omega}}{2\pi} \exp(-U_0/kT) \quad , \tag{4.27}$$

where for medium and strong damping, the prefactor $\widetilde{\omega}$ agrees with Kramers' result $(\widetilde{\omega} \cong (\omega_0^2 + \gamma^2)^{1/2} - \gamma, \ \gamma \equiv \eta/2M$, for a cubic potential), but for weak damping reduces not to his result (which is different from the above and tends to zero in the limit of zero damping) but rather to the so-called "simple transition state theory" prefactor $(\sim \omega_0)$. Since there seems no good reason to believe the transition state result for the steady-state (as distinct from initial) decay rate, it seems that these calculations cannot be the whole truth. Nevertheless, since there is no disagreement with the Kramer's result as regards the exponent[27] in Eq. (4.27), it seems probable that they do give this correctly and that the only question is about the prefactor. I will therefore summarize briefly the principal features of the results of these calculations.[28]

In the first place, one does indeed get a qualitative change in behavior at the temperature T_0 defined by Eq. (4.26). For the most extensively-studied case, that of ohmic dissipation with damping constant $\gamma \equiv \eta/2M$ in a cubic potential (which we will discuss from now on unless otherwise stated) one finds (Larkin and Ovchinnikov, 1984; Hanggi *et al.*, 1985)

$$\omega_m = (\omega_0^2 + \gamma^2)^{1/2} - \gamma \quad . \tag{4.28}$$

Hence for weak damping T_0 is approximately $\hbar\omega_0/2\pi k_B$, while for strong damping it is approximately $\hbar\omega_{\text{eff}}/2\pi k_B$, where $\omega_{\text{eff}} \equiv \omega_0^2/2\gamma$ is the characteristic (slow) frequency of classical motion in this limit. Above T_0 one essentially gets classical (Kramers) activation with some corrections due to the incipient quantum effects, while below T_0 (except in a narrow region very close to it) one is essentially dealing with quantum tunneling out of the ground-state and excited states of the metastable well. Well above T_0 the transition rate is given by the familiar "Kramers" formula

$$\Gamma = A \exp(-U_0/k_B T) \quad , \tag{4.29}$$

where the prefactor A as it results from these calculations is of order $\omega_0^2/2\gamma$ for large damping and ω_0 for small damping (but see above). The first quantum correction to the Kramers exponent can also be calculated (Hanggi *et al.*, 1985),

---

[27] That is, the term which is proportional to $\exp(-U_0/(\text{something}))$.

[28] For a more detailed review of work in this area, see Hanggi (1986).

and has the form

$$U_0/k_B T \rightarrow U_0/k_B T - \frac{1}{12}\left(\frac{\hbar\omega_0}{kT}\right)^2 \quad . \tag{4.30}$$

In view of the fact that the transition to quantum tunneling occurs, even for the undamped system, at $kT = \hbar\omega_0/2\pi$ rather than $kT = \hbar\omega_0$, these corrections can be quite important even in the "classical" regime $T > T_0$, and in fact there is an appreciable transition range where the simple formulas breakdown; the behavior in this region is discussed in detail in Grabert and Weiss (1984b) and Larkin and Ovchinnikov (1984).

Below T_0 the temperature-dependence of the exponent is much weaker. For the case of strong damping both the exponent and the prefactor in the WKB expression (Eq. (4.10)) can be evaluated analytically. The result is (Larkin and Ovchinnikov, 1984)

$$B = \left(\frac{2\pi}{9}\right)\eta(\Delta q)^2 \left(1 - \frac{1}{3}\left(\frac{T}{T_0}\right)^2\right) \quad , \tag{4.31}$$

$$A = \frac{2^{3/2}}{81}\eta^{7/2}\frac{(\Delta q)^3}{U_0 M^2} \quad . \tag{4.32}$$

Note that the exponent is independent of the mass for all temperatures below T_0 in this limit, while the prefactor is independent of temperature but depends on the mass. In the more general case one must resort to numerical methods to evaluate A and B, although for one special (T-dependent) value of $\alpha \equiv \gamma/\omega_0$ an analytical expression (Riseborough et al., 1985) has been obtained. In general, one finds that the more weakly damped the system, the less the temperature-dependence of the tunneling rate below T_0. One general statement can be made (Grabert et al., 1984; Grabert and Weiss, 1984b); for ohmic dissipation the first temperature corrections to the zero-temperature value of Γ are always of the form

$$\Gamma(T) = \Gamma(0)\exp(AT^2(M\omega_0 q_0^2/\hbar)) \quad , \tag{4.33}$$

where the constant A is related to the "length" of the zero-temperature bounce

$$\tau_B \equiv (\Delta q)^{-1} \int_{-\infty}^{\infty} q(t)dt \quad , \tag{4.34}$$

by the relation

$$A = \frac{\pi\gamma}{3\omega_0}(k_B\tau_B/\hbar)^2 \quad . \tag{4.35}$$

More generally, if $J(\omega) \propto \omega^n$ in the limit $\omega \to 0$ the first corrections to the exponent are of the form const. T^{n+1} .

The above predictions may be compared with the existing data (De Bruyn and Bol, 1982; Jackel *et al.*, 1981, Voss and Webb, 1981) on macroscopic quantum tunneling in current-based junctions and in SQUIDS. At the time of writing (May 1985), a new spate of data is just coming out (cf. below), and I will not attempt to review it here. However, it is probably fair to comment that the general order of magnitude of the measured tunneling rates and also the qualitative features of their dependences on the parameters (temperature, resistance, etc.) seem to be roughly as expected, which gives us some confidence that the general pattern of the theory described here is correct. Ironically, at the time of writing, the main difficulty seems to lie less in the quantum-mechanical calculation of the rates in terms of known parameters than in the extraction of the relevant parameters for a given experiment from the data on the quasi-classical motion (e.g., for a current-biased junction, from the current-voltage characteristics). I believe that the recent spate of experiments[29] is likely to stimulate new theoretical efforts in this direction; all we can say at present is that the agreement between theory and experiment is sufficient to encourage us to use the same models to consider the more spectacular type of experiment, that on macroscopic quantum coherence.

5. Macroscopic Quantum Coherence

The words "macroscopic quantum coherence" (MQC) are used to refer to an attempt to duplicate, at the level of a macroscopic coordinate, the kind of oscillatory behavior typical of an ammonia molecules or other kinds of microscopic system (e.g., the neutral K-meson) which have available to them two distinct but nearly degenerate states. To carry out such experiments one needs a macroscopic variable q (e.g., the flux trapped in an rf SQUID ring, see below) with an associated potential energy $V(q)$ which has two degenerate or nearly degenerate potential minima (see Fig. 6). One also needs a "kinetic energy", which as usual we write as $p^2/2M$ or $\frac{1}{2}M\dot{q}^2$, in order to ensure that the dynamics

---

[29] By the time this went to press (Dec. 1985) the spate of experimental papers referred to above had appeared in the literature: Washburn *et al.* (1985), Schwartz *et al.* (1985), Devoret *et al.* (1985), Martinis *et al.* (1985). In addition, a number of theoretical developments had occurred: Larkin and Ovchinnikov (1985, 1986), Ivlev and Mel'nikov (1985). Note in particular that Larkin and Ovchinnikov (1985) does obtain the Kramers weak-damping expression in the limit $T \gg T_0$. However, since the method used is different from the one described above, this does not completely resolve the questions raised concerning the latter.

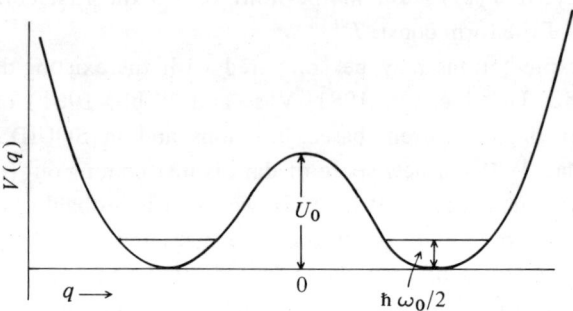

Fig. 6. Macroscopic quantum coherence.

is nontrivial. The quantity $[M^{-1} (\partial^2 V/\partial q^2)]^{1/2}$ evaluated at the minimum is denoted ω_0, and the barrier height U_0. In the ensuing discussion we shall always assume (a) that the barrier height U_0 is large enough compared to the zero-point energy $\hbar\omega_0/2$ in each well separately that the WKB approximation is applicable to the motion under the barrier, (b) that kT is very small compared to $\hbar\omega_0$. This has the double advantage that population of all but the two lowest-energy states of the double-well system is completely negligible, and that classical ("Kramers") processes of transition above the barrier are negligible. Let us first examine the quantum-mechanical predictions for this system under the assumption that it is totally isolated. (We will assume until further notice that the potential $V(q)$ is exactly symmetric around $q = 0$, and hence *a fortiori* that the two potential minima are equal.)

If we could neglect the possibility of tunneling through the barrier (e.g., take the limit $U_0 \to \infty$) then the quantum-mechanical state of the system would, trivially, be one of the states ψ_L, ψ_R corresponding to the groundstate in the left or right well; these would be approximately harmonic-oscillator states centered at $\pm q_0/2$, and would be degenerate, with an energy of approximately $\frac{1}{2}\hbar\omega_0$ relative to the bottom of the well. However, for any finite value of $U_0/\hbar\omega_0$ there exists a finite amplitude for tunneling between these states; this amplitude, which we denote $\frac{1}{2}\Delta$, is given by the formula

$$\frac{1}{2}\Delta = \text{const. } \omega_0 \exp\left(-\int (2MV(q))^{1/2} dq/\hbar\right), \tag{5.1}$$

which in general is of order $\omega_0 \exp(-\text{const. } U_0/\hbar\omega_0)$. (The specific formula for the realistic (SQUID ring) case is given in Eq. (2.40) in terms of the ring parameters.) As a result of the coherent tunneling between the two wells, the degeneracy of the groundstate is lifted and one gets a doublet of states:

$$\psi_+ = 2^{-1/2} \left(\psi_L + \psi_R \right) , \qquad E_+ = \frac{\hbar\omega_0}{2} - \frac{\Delta}{2} , \qquad (5.2a)$$

$$\psi_- = 2^{-1/2} \left(\psi_L - \psi_R \right) , \qquad E_- = \frac{\hbar\omega_0}{2} + \frac{\Delta}{2} . \qquad (5.2b)$$

Suppose, now, that we start the system at time zero in the left-hand well. It is then in a *linear combination* of the states ψ_+ and ψ_-, which have different energies. As a result, the system will oscillate between the two wells, in such a way that at times which are integral multiples of $2\pi/\Delta$ it will be found with certainty in the left well, and at half-odd-integral multiples of $2\pi/\Delta$ with certainty in the right well, while for all other time there is a finite probability to find it in either well. In fact, if we define $P(t)$ as the probability to find it in the left well minus the probability to find it in the right one, then we have the simple formula

$$P(t) = \cos(\Delta t) . \qquad (5.3)$$

It should be emphasized that although the time-dependence of $P(t)$ is exactly that of a simple harmonic oscillator, $P(t)$ in no sense represents the "actual position" of the system; the probability of actually finding it, on measurement, anywhere *between* the right well and the left one (i.e., under the barrier), is essentially zero (to be precise, it is of order $\exp(-U_0/\hbar\omega_0)$).

Let us suppose that we could conduct the relevant experiment on a system of the above type, when the minima $\pm q_0/2$ correspond to states which are macroscopically distinct by some agreed criterion. That is, we would start the system in the left-hand well at time zero, wait a time t without disturbing it, then measure the value of q (obtaining, we assume, the value $\pm q_0$ in all but an exponentially small fraction of cases); we would repeat this experiment many times, assigning in each case a value $+1$ if q is measured to be $+q_0/2$ and -1 if it is measured to be $-q_0/2$, and then take the average to obtain an "experimental" value of $P(t)$. By varying t we would get the complete "experimental" form of $P(t)$. Suppose that we could perform this experiment, and that the resulting "experimental" form of $P(t)$ agreed with the theoretical prediction in Eq. (5.3). What would be remarkable about this?

Let us first interpret the result within the quantum-mechanical formalism itself. Consider, for example, the state of the system at $t = \pi/(2\Delta)$. At that time, $P(t)$ is predicted to be zero, so there is an equal probability of finding the system in either of the two wells. Suppose now that we believed (as most formulations of the Copenhagen interpretation would seem to imply, cf. Sec. 1) that a macroscopic object cannot be in a linear superposition of macroscopically distinct states. In that case, we would be forced to assume that the true state of the system *is* at this time either ψ_L or ψ_R, with a probability of 50% each. But if we now take this as an initial condition and work out the subsequent development of the density matrix according to Schrodinger's equation, we get the trivial result that $P(t) = 0$ for all subsequent times — while Eq. (5.3) indicates, for example, that at $t = \pi/\Delta$, $P(t)$ is actually -1. Thus, if interpreted within the general quantum formalism, Eq. (5.3) *excludes* the possibility that the correct description of the system at time $\pi/2\Delta$ was a "mixture" of the states ψ_L and ψ_R, and confirms that it was indeed a linear superposition of these macroscopically distinct states.

This result is not particularly surprising — we should not expect to be able to preserve all the details of the quantum-mechanical predictions if we arbitrarily tinker with a particular bit of the quantum formalism which we happen to dislike, while leaving the rest intact. What is rather more remarkable is that Eq. (5.3), if confirmed experimentally, could be used to exclude a whole class of theories concerning the behavior of Nature at the macroscopic level, *without any reference to the quantum formalism.* It may be helpful to clarify this point by reference to the very similar situation which arises with respect to another classic paradox in the foundations of quantum mechanics, the EPR paradox. As is well known, in certain situations the standard quantum-mechanical description of two widely separated micro-systems 1 and 2 does not allow them separately to have any definite properties but rather describes the coupled system by a linear superposition of correlated states; e.g., if $|\uparrow\rangle_1$ indicates a state in which particle 1 has spin up, etc., then the wave function may of the form $2^{-1/2} (|\uparrow\rangle_1 |\downarrow\rangle_2 - |\downarrow\rangle_1 |\uparrow\rangle_2)$). Now, it was noted (Furry, 1936) very soon after Einstein *et al.*'s origin formulation (Einstein et. al., 1935) of their paradox that the assumption that the correct description of the coupled system was not of the above linear-superposition form, but rather of the form of a mixture of $|\uparrow\rangle_1 |\downarrow\rangle_2$ and $|\downarrow\rangle_1 |\uparrow\rangle_2$, would — *provided that this mixture, and the results of measurements on it, were interpreted in the standard quantum-mechanical way* — give predictions different from those of the standard description; and some experiments were done which verified the standard prediction and therefore excluded the "mixture" hypothesis (stage I). However, nearly thirty years later Bell proved a very much stronger theorem, (Bell, 1964), namely that under certain appropriate conditions, an experimental result which agreed with the standard quantum-mechanical prediction must automatically exclude a *whole class* of theories about the description of widely separated microscopic systems (the class of so-called "objective local" theories) *quite without reference to any quantum-mechanical considerations.* Since the appearance of Bell's work, a series of experiments has been done which confirm the quantum predictions in the appropriate regime and hence — at least in the opinion of most physicists — decisively exclude such objective local theories (stage II). The considerations of the last paragraph correspond in the MQC context to stage I in the history of the EPR paradox; those we shall now describe correspond to stage II.

Consider (Leggett and Garg, 1985) the following pair of assumptions about reality at the macroscopic level: (A1) A macroscopic object which has available to it two (or more) macroscopically distinct states must at all[30] times "actually be" in one or the other of these states; (A2) It is possible, at least in principle, to verify that the macroscopic object is in a definite macroscopic state with arbitrarily small effect on its subsequent behavior. These assumptions seem to be part of what we would (in our everyday life outside the physics laboratory) think of as common sense. The surprising thing is that, in the context of an idealized MQC experiment, the conjunction of (A1) and (A2) is incompatible with Eq. (5.3)! Probably the simplest way to see this is the following. Define a quantity $Q(t)$ which has the value $+1$ (-1) if at time t the system is in the left (right) well. According to assumption (A1), for all times t, $Q(t)$ "actually has" the value ± 1. Consider an "experiment" in which the system is started in a definite state at time zero and the left unobserved. It is trivial to establish (e.g., by exhaustion of the 16 possibilities for the values of the $Q(t_i)$) the inequality

$$Q(t_1)Q(t_2) + Q(t_2)Q(t_3) + Q(t_3)Q(t_4) - Q(t_1)Q(t_4) \leqslant 2 .$$

(5.4)

Suppose, now, that it were somehow possible to measure the expectation values of all of the four quantities on the left-hand side of Eq. (5.4) on the same ensemble. Then, putting $\langle Q(t_1)Q(t_2) \rangle \equiv K_{12}$, etc., where the brackets indicated the experimentally measured expectation value, we should predict

$$K_{12} + K_{23} + K_{34} - K_{14} \leqslant 2 .$$

(5.5)

By contrast, the prediction of quantum mechanics for an arbitrary ensemble which is undisturbed between t_i and t_j is simply that $K_{ij} \equiv P(t_j - t_i)$, where $P(t)$ is the function given by Eq. (5.3).[31] It is easily verified that for certain values of the t_i (e.g., $t_2 - t_1 = t_3 - t_2 = t_4 - t_3 = \pi\Delta/4$) the inequality (5.5) is violated.

[30] More accurately, "almost all". See Leggett and Garg (1985).

[31] This follows because an initial measurement of $Q(t_i)$ which gives ± 1 imposes the boundary condition $\psi(t_i) = \psi_L$ or ψ_R and because, if the initial state is ψ_R, $P(t)$ is of course just $-\cos \Delta t$.

Unfortunately, it is impossible to conclude without further argument that we could discriminate experimentally between quantum mechanics and a theory possessing the properties (A1) and (A2). The reason is quite subtle. The quantum-mechanical predictions for the K_{ij} apply only to ensemble which are undisturbed between times t_i and t_j. Thus, in conducting the experiment one cannot make the measurements of all of the $Q(t_i)$ in succession on a single run (any such measurement would, of course, trivially satisfy Eq. (5.4); rather one has to organize a series of runs in which one starts the system off in a well-defined state at (say) t_1 and measures the value of $Q(t_2)$, then a second series of runs in which once again starts in the same state as t_1, but now does *not* measure $Q(t_2)$ but rather $Q(t_3)$, and so on. The problem is that the ensemble corresponding to the second group of runs is not the same as that corresponding to the first, since in the former case, but not the latter, a measurement was made at t_2. This is where assumption (A2) comes in; it allows one to assert that it is possible, in principle at least, to make the measurement in such a way that (in a theory satisfying (A1)) the properties of the ensemble are not affected by the measurement. Thus, one can now assert that the inequality (5.5) should hold at least when the experimental values are measured in this "ideal" way; and one then only needs the rather weak additional assumption that the values of the individual K_{ij}, if they could be measured in this ideal way, would be identical to those measured in a more practical (non-ideal) way.[32] Under these conditions, then, one concludes that any theory satisfying the constraints (A1) and (A2) should predict values of the *experimentally measured* quantities of K_{ij} which satisfy the inequality (5.5). Since, as we have seen, for certain choices of the t_i the predictions of quantum mechanics violate (5.5), we now have a clear means to discriminate between quantum mechanics and a whole class of alternative "macro-realistic" theories.

So far, so good. However, the quantum-mechanical prediction in Eq. (5.3) was made for an isolated system, and much of the discussion of the previous sections has focused precisely on the fact that macroscopic systems are *not* isolated from their environments. How does this modify the above argument? It is clear that it makes absolutely no difference to the conclusion that the inequality (5.5) holds for macrorealistic theories, since the argument leading to that conclusion was a purely algebraic one. The only difference is that we should in general expect that the quantum-mechanically predicted values of the quantities K_{ij} would be modified by the interaction of the system with its environment. Such a modification is, indeed, qualitatively predicted by

---

[32] This assumption seems eminently reasonable, since the ideality or otherwise of the measurement process should affect only what goes on *after* the measurement.

simple arguments from quantum measurement theory (Simonius, 1978); in fact, it is usually believed that in the presence of a substantial system-environment interaction the behavior of $P(t)$ (and hence, as it turns out, of the $K(t_j - t_i)$) changes qualitatively from (5.3) to the form

$$K_{ij} = P(t_j - t_i), \quad P(t) \sim e^{-\gamma t} , \tag{5.6}$$

which automatically satisfies the inequality (5.5) for any choice of the t_i. Our task, then, is to verify (or not) this assertion by a calculation as concrete as possible.

While it is, obviously, impossible to obtain conclusions which are totally independent of the description of the system-environment interaction, once we have agreed to model the latter by the Hamiltonian (3.7) we can do a quantitative calculation. In fact, if we start from (3.7), with the potential energy $V(q)$ having the shape appropriate to an MQC experiment (cf. Fig. 6) we can actually calculate $P(t)$ quantitatively, and hence infer[33] the values of the K_{ij}. This calculation is carried out in detail in Leggett *et al.* (1986), and I merely survey the results here. The principal results are as follows. (1) Irrespective of the detailed form of the spectral density $J(\omega)$ of the system-environment coupling Eq. (3.9) (or equivalently of the form of the quantity $K(\omega)$, Eq. (3.10)), the interaction renormalizes the effective tunneling frequency Δ downwards to a value Δr, often by a very large factor. (2) If $J(\omega)$ tends to zero as ω^n, as $\omega \to 0$, where $n < 1$, then at $T = 0$ the system is localized in the well in which it starts, i.e., $P(t) = 1$ for all t; at finite temperatures $P(t)$ satisfies Eq. (5.6), with a temperature-dependent rate constant γ. (3) On the contrary, if $J(\omega)$ behaves as ω^n, $n > 1$, then at $T = 0$ the system executes oscillations of the form (5.3), but with Δ replaced by the renormalized frequency Δ_r, and these oscillations are only weakly damped. (4) The case of most physical interest in the real-life (SQUID-ring) context is when $J(\omega)$ is linearly proportional to ω in the limit $\omega \to 0$; this corresponds to linear ohmic dissipation in the classical motion (that is, the equation of motion contains a term $\eta \dot{q}$, where η (R^{-1} in the SQUID case) is a friction coefficient). In this case we can define a dimensionless constant[34]

$$\alpha \equiv \eta q_0^2 / 2\pi \hbar , \tag{5.7}$$

---

[33] For a dissipative system we cannot strictly write the identity $K_{ij} = P(t_j - t_i)$ which holds in the isolated case. However, arguments are given in Leggett *et al.* (1986) to show that, in the "interesting" region of parameters the difference between the two quantities is small and does not affect the qualitative conclusions.

[34] The result is quoted explicitly for the case when (3.8) holds. For the more general case, see Leggett *et al.* (1986).

which in the SQUID case can be written $(\Delta\Phi/\Phi_0)^2$ (R_0/R), where $\Delta\Phi$ is the difference in trapped flux Φ between the two minimum of $U(\Phi)$ and R_0 is the "quantum unit of resistance", $h/4e^2$ ($\sim 6\,\mathrm{k}\Omega$). It then turns out that if we study the behavior in the parameter plane defined by α and the temperature T, over most of the plane the behavior is approximately that described by (5.6); in this sense we verify the qualitative predictions of the arguments from quantum measurement theory. However, there is small corner of the parameter space, defined roughly speaking, by the conditions

$$\alpha \ll 1 \ , \qquad \alpha k T/\Delta_r \ll 1 \ , \tag{5.8}$$

where the system shows weakly damped oscillations analogous to those predicted by (5.3). In addition, by going to small enough values of α and T one can obtain quantum-mechanical predictions which violate the inequality (5.5) (and other similar inequalities). Thus, provided the conditions (5.8) can be met simultaneously with the many other conditions necessary for an MQC experiment, it should be possible to force Nature, as it were, to choose between quantum-mechanics and realism at a macroscopic level. I believe (though this may perhaps be as yet minority view) that a clear-cut outcome of this experiment in either direction would be of profound significance for the discussion of the foundations of quantum mechanics.

Acknowledgments

This paper is dedicated to the memory of Professor Shang-keng Ma. It was supported through the MacArthur Professorship endowed by the John D. and Catherine T. MacArthur Foundation at the University of Illinois. I am grateful to Peter Hänggi for helpful comments on the manuscript.

References

Affleck, I. K. (1981) *Phys. Rev. Lett.* **46**, 388.

Ambegaokar, V. (1969) in *Superconductivity*, Parks, R. D., ed. (Marcel Dekker, New York), Vol. I.

Ambegaokar, V., Eckern, U. and Schon, G. (1982) *Phys. Rev. Lett.* **48**, 1745.

Anderson, P. W. (1964) in *The Many-Body Problem*, Caianello, E. R., ed. (Academic, New York and London), vol. 2.

Anderson, P. W. (1984) *Basic Notions of Condensed Matter Physics* (Benjamin-Cummings, Menlo Park), pp. 49-50.

Banks, T., Bender, C. M. and Wu, T. T. (1973) *Phys. Rev.* **D8**, 3346.

Bardeen, J., Cooper, L. N. and Schrieffer, J. R. (1957) *Phys. Rev.* **108**, 1175.

Bardeen, J. (1961) *Phys. Rev. Lett.* **6**, 57.

Bardeen, J. (1979) *Phys. Rev. Lett.* **42**, 1498.

Bell, J. S. (1964) *Physics* **1**, 195.

Bloch, F. (1970) *Phys. Rev.* **B2**, 109.

Bohr, N. (1963) *Essays 1958-62 on Atomic Physics and Human Knowledge* (Interscience, New York).

Bub, J. (1968) *Nuovo Cimento* **B73**, 503.

Buttiker, M. and Landauer, R. (1981) *Phys. Rev.* **B24**, 4079.

Buttiker, M. and Landauer, R. (1983) *Phys. Rev.* **B28**, 1268.

Caldeira, A. O. and Leggett, A. J. (1983) *Ann. Phys.* (New York) **149**, 374; *erratum* (1984) *ibid.* **153**, 445.

Callan, C. G. and Coleman, S. (1977) *Phys. Rev.* **D16**, 1762.

Chang, L.-D. and Chakravarty, S. (1984) *Phys. Rev.* **B29**, 130; *erratum* (1984) *ibid.* **B30**, 1566.

Chang, L.-D. and Chakravarty, S. (1985) *Phys. Rev.* **B31**, 154.

Chen, Y.-C., Fisher, M. P. A. and Leggett, A. J. (1986) (unpublished).

Coleman, S. (1979) in *The Whys of Subnuclear Physics* Zichichi, A., ed. (Plenum, New York).

De Bruyn Ouboter, R. and Bol, D. (1982) *Physica* **B15**, 112, and earlier work cited therein.

d'Espagnat, B. *(1976) Conceptual Foundations of Quantum Mechanics* 2nd edition (Benjamin, Reading, Mass) part IV.

De Vega, H. J., Gervais, J. L. and Sakita, B. (1976) *Phys. Rev.* **D19**, 604.

Devoret, M. H., Martinis, J. M. and Clarke, J. (1985) *Phys. Rev. Lett.* **55**, 1908.

de Witt, B. S. and Graham, R. N. (1971) *Am. J. Phys.* **39**, 724.

Dirac, P. A. M. (1958) *The principles of Quantum Mechanics* (Clarendon Press, Oxford).

Eckern, U., Schon, G. and Ambegaokar, V. (1984) *Phys. Rev.* **B30**, 6419.

Einstein, A., Podolsky, B. and Rosen, N. (1935) *Phys. Rev.* **47**, 777.

Feynman, R. P. and Hibbs, A. R. (1965) *Quantum Mechanics and Path Integrals* (McGraw-Hill, New York).

Furry, W. H. (1936) *Phys. Rev.* **49**, 393.

Gasiorowicz, S. (1966) *Elementary Particle Physics* (Wiley, New York), pp, 50-54.

Grabert, H., and Weiss, U. and Hanggi, P. (1984) *Phys. Rev. Lett.* **52**, 2193.

Grabert, H. and Weiss, U. (1984a) *Z. Phys.* **B56**, 171.

Grabert, H. and Weiss, U. (1984b) *Phys. Rev. Lett.* **53**, 1787.

Grabert, H., Olschowski, P. and Weiss, U. (1985) *Phys. Rev.* **B32**, 3348.

Hanggi, P. (1986) *J. Stat. Phys.* **42** 105.

Hanggi, P., Grabert, H., Ingold, G.-L. and Weiss, U. (1985) *Phys. Rev. Lett.* **55**, 761.

Hepp, K. (1972) *Helv. Phys. Acta* **45**, 237.

Ivlev, B. N. and Mel'nikov, V. I. (1985) *Phys. Rev. Lett.* **55**, 1614.

Jackel, L. D., Gordon, J. P., Hu, E. L., Howard, R. E., Fetter, L. A., Tennant, D. M., Epworth, R. and Kurkijarvi, J. (1981) *Phys. Rev. Lett.* **47**, 697.

Jaklevic, R. C., Lambe, J., Mercereau, J. E. and Silver, A. H. (1965) *Phys. Rev.* **A140**, 1628.

Jammer, M. (1974) *The Philosophy of Quantum Mechanics* (Wiley-Interscience, New York), pp. 206-7.

Joos, E. and Zeh, H. D. (1985) *Z. Phys.* **B59**, 223.

Josephson, B. D. (1965) *Adv. Phys.* **14**, 419.

Kapur, P. L. and Peierls, R. E. (1937) *Proc. Roy. Soc.* (London) **A163**, 606.

Kramers, H. A. (1940) *Physica* **7**, 284.

Landauer, R. (1950) thesis, Harvard University (unpublished).

Langer, J. S. (1967) *Ann. Phys.* (New York) **41**, 108.

Larkin, A. I. and Ovchinnikov, Yu. N. (1982) *Phys. Rev.* **B28**, 6281.

Larkin, A. I. and Ovchinnikov (1983a) *Pis'ma Zh. Eksp. Teor. Fiz.* **37**, 322 [(1983) *JETP Lett.* **37**, 382].

Larkin, A. I. and Ovchinnikov (1983b) *Zh. Eksp. Teor. Fiz.* **85**, 1510 [(1983) *Sov. Phys. JETP* **58**, 876].

Larkin, A. I. and Ovchinnikov, Yu. N. (1984) *Zh. Eksp. Teor. Fiz.* **86**, 719 [(1984) *Sov. Phys. JETP* **59**, 420].

Larkin, A. I. and Ovchinnikov, Yu. N. (1985) *J. Stat. Phys.* **41**, 425.

Larkin, A. I. and Ovchinnikov (1986b) *Resonance Reduction of Lifetime of Metastable States of Tunnel Junctions* (preprint).

Leggett, A. J. (1966) *Prog. Theor. Phys.* **36**, 901.

Leggett, A. J. (1980) *Prog. Theor. Phys. Suppl.* No. **69**, 80.

Leggett, A. J. (1984a) in *Proc. Intl. Symposium on the Foundations of Quantum Mechanics in the Light of New Technology*, Kamefuchi, S., *et al.*, eds., (Japanese Physical Society, Tokyo).

Leggett, A. J. (1984b) in *Percolation, Localization, and Superconductivity* (NATO ASI Series 8, Physics V. 109) Goldman, A. M. and Wolf, S. A., eds. (Plenum Press, New York).

Leggett, A. J. (1984c) *Phys. Rev.* **B30**, 1208.

Leggett, A. J. (1985) in *The Lesson of Quantum Theory,* de Boer, J., Dal, E., and Ulfbech, O., eds. (Proc. Niels Bohr Centennial Symposium, Copenhagen, Oct, 1985), North-Holland, Amsterdam, 1986.

Leggett, A. J., Chakravarty, S., Dorsey, A. T., Fisher, M. P. A., Garg, A. and Zwerger, W. (1986) *Rev. Mod. Phys.,* to be published.

Leggett, A. J. and Garg, A. (1985) *Phys. Rev. Lett.* **54**, 587.

Leggett, A. J. (1986) in *Quantum Theory and Beyond,* Hiley, B. J. and Pent, F. D., eds. (Routledge and Kegan Paul).

Lett, P., Christian, W., Singh, S. and Mandel, L. (1981) *Phys. Rev. Lett.* **47**, 1892.

Likharev, K. K. and Zorin, A. B. (1985) *J. Low Temp. Phys.* **59**, 347.

Lounasmaa, O. V. (1974) *Experimental Principles and Methods Below 1K* (Cademic Press, New York), Chap. 7.

Martinis, J. M., Devoret, M. H., and Clarke, J. (1985) *Phys. Rev. Lett.* **55**, 1543.

McCumber, D. E. (1968) *J. Appl. Phys.* **39**, 3113.

Mel'nikov, V. I. and Meshkov, S. V. (1983) *Pis'ma Zh. Eksp. Teor. Fiz.* **38**, 111 (1983) [*JETP* Letters **38**, 130].

Riseborough, P., Hanggi, P. and Freidkin, E. (1985) *Phys. Rev.* **A32**, 489.

Rogovin, D. and Nagel, J. (1982) *Phys. Rev.* **B26**, 3698.

Schafroth, M. R. (1955) *Phys. Rev.* **100**, 463.

Schmid, A. (1986) *Ann. Phys.* (in press).

Schrödinger, E. (1935) *Die Naturwissenschaften* **23**, 844.

Schwartz, D. B., Sen, B., Archie, C. N. and Lukens, J. E. (1985) *Phys. Rev. Lett.* **55**, 1547.

Simonius, M. (1978) *Phys. Rev. Lett.* **40**, 980.

Stewart, W. C. (1968) *Appl. Phys. Lett.* **12**, 277.

Stone, M. (1977) *Phys. Lett.* **B67**, 186.

Voss, R. F. and Webb, R. A. (1981) *Phys. Rev. Lett.* **47**, 265.

Washburn, S., Webb, R. A., Voss, R. F. and Faris, S. M. (1985) *Phys. Rev. Lett.* **54** 2172.

Waxman, D. (1984) D. Phil. thesis, University of Sussex (unpublished).

Waxman, D., and Leggett, A. J. (1985) *Phys. Rev.* **B32**, 4450.

Weiss, U., Riseborough, O., Hanggi, P. and Grabert, H. (1984) *Phys. Lett.* **104A**, 10; *erratum* (1984) **ibid. 104A**, 492.

Widom, A., Megaloudis, G., Sacco, J. E. and Clark, T. D. (1981) *Nuovo Cimento* **B61**, 112.

Wolynes, P. G. (1981) *Phys. Rev. Lett.* **44**, 968.

PUBLICATIONS OF SHANG-KENG MA

Books

1. Shang-keng Ma, *Modern Theory of Critical Phenomena* (Benjamin, Reading, Mass., 1976).
2. Shang-keng Ma, *Statistical Mechanics* (World Scientific, Singapore, 1985).

Articles

1. Shang-keng Ma and Chia-Wei Woo, "Theory of a Charged Bose Gas. I," *Physical Review* **159** (1967) 165–175.
2. Chia-Wei Woo and Shang-keng Ma, "Theory of a Charged Bose Gas. II," *Physical Review* **159** (1967) 176–183.
3. Shang-keng Ma and Keith A. Brueckner, "Correlation Energy of an Electron Gas with a Slowly Varying High Density," *Physical Review* **165** (1968) 18–31.
4. Shang-keng Ma, M. T. Beal-Monod and D. R. Fredkin, "Temperature Dependence of the Spin Susceptibility of a Nearly Ferromagnetic Fermi Liquid," *Physical Review Letters* **20** (1968) 929–932.
5. Shang-keng Ma and H. Gould, "Low Temperature Mobility of Heavy Impurities in Fermi Liquids," *Physical Review Letters* **21** (1968) 1379–1382.
6. Shang-keng Ma, M. T. Beal-Monod and D. R. Fredkin, "Spin Waves in He^3 in the Paramagnon Model," *Physical Review* **174** (1968) 227–239.
7. Shang-keng Ma and S. J. Chang, "Feynman Rules and Quantum Electrodynamics at Infinite Momentum," *Physical Review* **180** (1969) 1506–1513.
8. Shang-keng Ma and S. J. Chang, "Scattering Amplitudes in Quantum Electrodynamics at Infinite Energy," *Physical Review Letters* **22** (1969) 1334–1337.
9. Shang-keng Ma and H. Gould, "Low Temperature Ion Mobility in Interacting Fermi Liquids," *Physical Review* **183** (1969) 338–348.
10. Shang-keng Ma, R. Dashen and H. J. Bernstein, "S-Matrix Formulation of Statistical Mechanics," *Physical Review* **187** (1969) 345–370.
11. Shang-keng Ma and S. J. Chang, "Multiphoton Exchange Amplitudes at Infinite Energy," *Physical Review* **188** (1969) 2385–2404.
12. Shang-keng Ma and R. Dashen, "Singular Three-Body Amplitudes in the Theory of the Third Virial Coefficient," *Journal of Mathematical Physics* **11** (1970) 1136–1143.

13. Shang-keng Ma, H. Gould and V. K. Wong, "Phonon Dispersion in a Low Temperature Weakly Interacting Bose Gas," *Physical Review* **A3** (1971) 1453–1462.

14. R. Dashen and Shang-keng Ma, "Singularities in Forward Multiparticle Scattering Amplitudes and the S-Matrix Interpretation of Higher Virial Coefficients," *Journal of Mathematical Physics* **12** (1971) 689–715.

15. R. Dashen and Shang-keng Ma, "S-Matrix and Low Energy Theorem in the Theory of Correlation Functions," *Journal of Mathematical Physics* **12** (1971) 1449–1471.

16. R. Dashen and Shang-keng Ma, "Scattering Theory and Current Correlations in Classical Gases," *Physical Review* **A4** (1971) 700–707.

17. Shang-keng Ma, "Second Sound in a Low Temperature Weakly Interacting Bose Gas," *Journal of Mathematical Physics* **12** (1971) 2157–2170.

18. Shang-keng Ma, "Transverse Waves in a Weakly Interacting Bose Gas," *Physical Review* **A5** (1972) 2632–2642.

19. Shang-keng Ma, "Critical Exponents for Charged and Neutral Bose Gases Above λ-Points," *Physical Review Letters* **29** (1972) 1311–1314.

20. M. E. Fisher, Shang-keng Ma and B. G. Nickel, "Critical Exponents for Long-Range Interactions," *Physical Review Letters* **29** (1972) 917–920.

21. Shang-keng Ma, "Critical Exponents above T_c to $O(1/n)$," *Physical Review* **A7** (1973) 2172–2187.

22. B. I. Halperin, P. C. Hohenberg and Shang-keng Ma, "Calculation of Dynamic Critical Properties using Wilson's Expansion Methods," *Physical Review Letters* **29** (1972) 1548–1551.

23. Shang-keng Ma, "Renormalization Group in the Large n Limit," *Physics Letters* **43A** (1973) 475–476.

24. Shang-keng Ma, "Renormalization Group and the Large n Limit," *Journal of Mathematical Physics* **15** (1974) 1866–1891.

25. Shang-keng Ma, "Introduction to the Renormalization Group," *Reviews of Modern Physics* **45** (1973) 589–614.

26. B. I. Halperin, T. C. Lubensky and Shang-keng Ma, "First-Order Phase Transitions in Superconductors," *Physical Review Letters* **32** (1974) 292–295.

27. Shang-keng Ma, "Scaling Variables and Dimensions," *Physical Review* **A10** (1974) 1818–1836.

28. B. I. Haperin, P. C. Hohenberg and Shang-keng Ma, "Renormalization Group Methods for Critical Dynamics. I: Recursion Relations and Effects of Energy Conservation," *Physical Review* **B10** (1974) 139–153.

29. Shang-keng Ma and L. Senbetu, "Sound Waves near T_c in a Dynamic Spherical Model," *Physical Review* **A10** (1974) 2401–2422.

30. R. Dashen, Shang-keng Ma and R. Rajaraman, "Finite Temperature Behavior of a Relativistic Field Theory with Dynamical Symmetry Breaking," *Physical Review* **D11** (1975) 1499–1508.

31. Shang-keng Ma and R. Rajaraman, "Comments on the Absence of Spontaneous Symmetry Breaking in Low Dimensions," *Physical Review* **D11** (1975) 1701–1704.

32. Shang-keng Ma and G. F. Mazenko, "Critical Dynamics of Ferromagnets in 6-ε Dimensions," *Physical Review Letters* **33** (1974) 1383–1385.
33. Shang-keng Ma and G. F. Mazenko, "Critical Dynamics of Ferromagnets in 6-ε Dimensions: General Discussion and Detailed Calculation," *Physical Review* **B11** (1975) 4077–4100.
34. Shang-keng Ma, "The $1/n$ Expansion," in *Phase Transitions and Critical Phenomena*, eds. C. Domb and M. S. Green (Academic Press, New York, 1976).
35. A. Aharony, Y. Imry and Shang-keng Ma, "Comments on the Critical Behavior of Random Systems," *Physical Review* **B13** (1976) 466–473.
36. Y. Imry and Shang-keng Ma, "Random Field Instability of the Ordered State of Continuous Symmetry," *Physical Review Letters* **35** (1975) 1399–1401.
37. B. I. Halperin, P. C. Hohenberg and Shang-keng Ma, "Renormalization Group and Methods for Critical Dynamics. II: Detailed Analysis of the Relaxational Models," *Physical Review* **B13** (1976) 4119–4131.
38. Shang-keng Ma, "Scale Transformations in Dynamic Models," Lecture Notes in Physics, eds. J. Ehlers *et al.*, in *Critical Phenomena. Sitges International School on Statistical Mechanics*, eds. J. Brey and R. B. Jones, Vol. 54 (Springer-Verlag, Berlin, 1976).
39. Shang-keng Ma, "Renormalization Group by Monte Carlo Methods," *Physical Review Letters* **37** (1976) 461–464.
40. G. Grinstein, Shang-keng Ma and G. F. Mazenko, "Dynamics of Spins Interacting with Quenched Random Impurities," *Physical Review* **B15** (1977) 258–272.
41. A. Aharony, Y. Imry and Shang-keng Ma, "Lowering of Dimensionality in Phase Transitions with Random Fields," *Physical Review Letters* **37** (1976) 1364–1367.
42. C. De Dominicis, Shang-keng Ma and L. Peliti, "Critical Dynamics Near Dimension Two for Time-Dependent Ginsburg-Landau Models," *Physical Review* **B15** (1977) 4313–4317.
43. Shang-keng Ma and J. Rudnick, "Time-Dependent Ginzburg-Landau Model of the Spin Glass," *Physical Review Letters* **40** (1978) 589–593.
44. Shang-keng Ma, "Effect of Random Impurities on Long Range Order," *Chinese Journal of Physics* **16** (1978) 146–152.
45. Shang-keng Ma, "Comments on Recent Advances in Theoretical Physics," *Science and Technology* **2** (1978) 4–10 (in Chinese).
46. Shang-keng Ma, "Alternative Approach to the Dynamic Renormalization Group," *Physical Review* **B19** (1979) 4824–4837.
47. Chandan Dasgupta, Shang-keng Ma and Chin-Kun Hu, "Dynamic Properties of a Spin-Glass Model at Low Temperatures," *Physical Review* **B20** (1979) 3837–3849.
48. Shang-keng Ma, Chandan Dasgupta and Chin-kun Hu, "Random Antiferromagnetic Chain," *Physical Review Letters* **43** (1979) 1434–1437.
49. Chandan Dasgupta and Shang-keng Ma, "Low-temperature Properties of the Random Heisenberg Antiferromagnetic Chain," *Physical Review* **B22** (1980) 1305–1319.

50. Shang-keng Ma, "Dynamics of a Vector Spin-Glass Model," *Physical Review* **B22** (1980) 4484–4502.

51. Daniel J. Amit, Shang-keng Ma and R. K. P. Zia, "The $O(n)$-Symmetric Model Between Two and Four Dimensions," *Nuclear Physics* **B180** (1981) 157–162.

52. Shang-keng Ma, "Calculation of Entropy from Data of Motion," *Journal of Statistical Physics* **26** (1981) 221–240.

53. Shang-keng Ma and Matthew Payne, "Entropy of a Spin-Glass Model with a Long-Range Interaction," *Physical Review* **B24** (1981) 3984–3990.

54. Hsin-Hsiung Chen and Shang-keng Ma, "Low Temperature Behavior of a One-Dimensional Random Ising Model," *Journal of Statistical Physics* **29** (1982) 717–746.

55. G. Grinstein and Shang-keng Ma, "Roughening and Lower Critical Dimension in the Random-Field Ising Model," *Physical Review Letters* **49** (1982) 685–688.

56. Shang-keng Ma, "One-Dimensional Boltzmann Equation with a Three-Body Collision Term," *Journal of Statistical Physics* **31** (1983) 107–114.

57. G. Grinstein and Shang-keng Ma, "Surface Tension, Roughening and Lower Critical Dimension in the Random-Field Ising Model," *Physical Review* **B28** (1983) 2588–2601.